Statistics and Probability Theory

TOPICS IN SAFETY, RISK, RELIABILITY AND QUALITY

Volume 18

Aims and Scope. Fundamental questions which are being asked these days of all products, processes and services with ever increasing frequency are:

What is the risk?
How safe is it?
How reliable is it?
How good is the quality?
How much does it cost?

This is particularly true as the government, industry, public, customers and society become increasingly informed and articulate.

In practice none of the three topics can be considered in isolation as they all interact and interrelate in very complex and subtle ways and require a range of disciplines for their description and application; they encompass the social, engineering and physical sciences and quantitative disciplines including mathematics, probability theory and statistics.

The major objective of the series is to provide series of authoritative texts suitable for academic taught courses, reference purposes, postgraduate and other research and practitioners generally working or strongly associated with areas such as:

Safety Assessment and Management
Emergency Planning
Risk Management
Reliability Analysis and Assessment
Vulnerability Assessment and Management
Quality Assurance and Management

Special emphasis is placed on texts with regard to readability, relevance, clarity, applicability, rigour and generally sound quantitative content.

For further volumes:
www.springer.com/series/6653

Michael Havbro Faber

Statistics and Probability Theory

In Pursuit of Engineering Decision Support

Springer

Michael Havbro Faber
Civil Engineering dept.
DTU – Danish Technical University
Kgs. Lyngby, Denmark

Additional material to this book can be downloaded from http://extras.springer.com.

ISSN 1566-0443 Topics in Safety, Risk, Reliability and Quality
ISBN 978-94-017-8530-3 ISBN 978-94-007-4056-3 (eBook)
DOI 10.1007/978-94-007-4056-3
Springer Dordrecht Heidelberg London New York

Printed on acid-free paper

Springer is part of Springer Science+Business Media (www.springer.com)

This book is dedicated to the ones I love

Preface

Decisions in general—good ones and bad ones—shape our world and direct our lives. The success of the world society at local and global scale depends on its ability to make the right decisions.

Engineers in particular, play an important role for society as they are directly involved in establishing the basis for societal decision making.

The value of decisions is and will always be relative to the objectives as well as the assumed boundary conditions. From a theoretical perspective, if we know what we want and if we know how to get there, the task is straight forward. In reality, however, the situation is quite different, we might be able to agree on what we want but we are rarely sure on how to get there. The reason for this is simple but challenging, we do not exactly know what the consequences of different decision alternatives are because we do not completely understand the world we are living in. Our knowledge is limited and imprecise, and we make errors.

Statistics and probability theory is often presented to students in general, and in engineering in particular, as a topic of value in itself and for this reason it is seldom really appreciated for what it can be used for, namely as a basis for assessing information and organizing the process of acquiring knowledge in pursuit of supporting decision making.

This book has exactly this aim, namely to present statistics and probability theory in the context of supporting engineering decision making. It is assumed that the reader has only little or no prior knowledge on the subject of statistics and probability theory, why this book would be appropriate for undergraduate engineering students. Moreover, unlike many standard textbooks on the same subject, the perspective of the present book is to focus on the use of the theory for the purpose of engineering model building and decision making.

The book is subdivided into the following seven chapters, each consisting of one or more lectures as fitting for a typical semester (see also Fig. 1):

Chapter 1 Engineering decisions under uncertainty
Chapter 2 Basic probability theory
Chapter 3 Descriptive statistics
Chapter 4 Uncertainty modeling

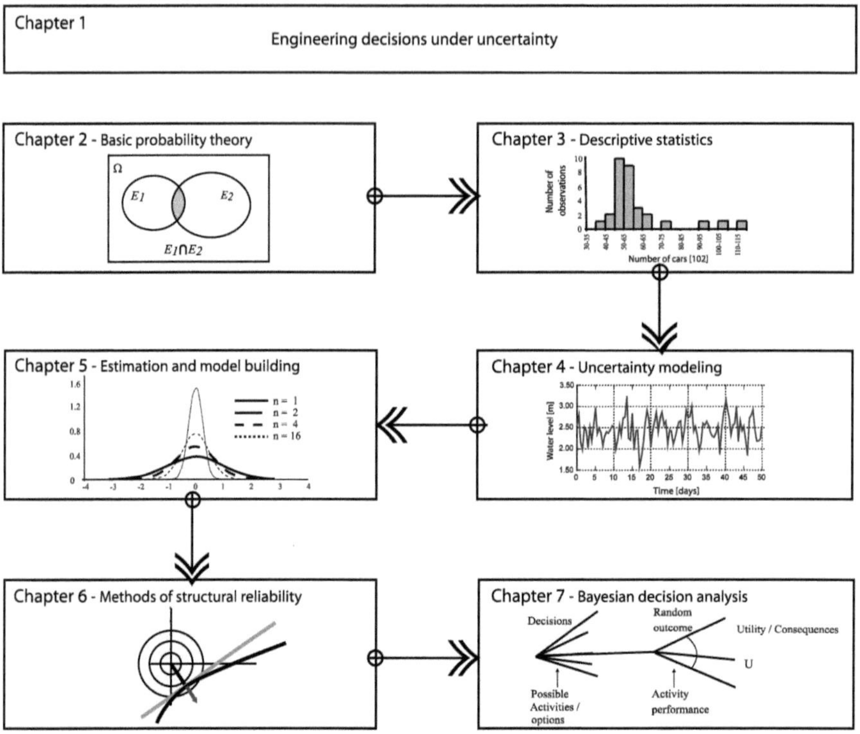

Fig. 1 Illustration of the modules of the course and their didactical roles

Chapter 5 Estimation and model building
Chapter 6 Methods of structural reliability
Chapter 7 Bayesian decision analysis

Chapter 1 first provides a motivation for the application of statistics and probability as a basis for developing engineering models and for risk based decision making. Thereafter, Chap. 2 introduces the basic theory of probability. Subsequently, in Chap. 3 a selection of tools is provided, which enables engineers to assess and communicate data in a condensed form, namely descriptive statistics. In Chap. 4 an introduction to uncertainty is provided together with a description of the various building stones required to represent uncertainties in engineering modeling in terms of random variables and processes. In Chap. 5 the main focus is directed to the aspects of postulating models, assessing model parameters and verifying models. Chapter 6 shows how, on the basis of formulated probabilistic models of uncertain variables, probabilities of events typical in engineering applications may be assessed. Finally, in Chap. 7 it is shown how the engineering models of uncertainties and their probabilistic descriptions can be utilized in a systematic framework for engineering decision making.

It is believed that the reader who has completed the present script, will be able to:

- assess data based on observations and/or experiment results and present these in a standardized and unambiguous form,
- formulate and validate simple engineering models with due consideration of the associated uncertainties due to lack of knowledge and data as well as natural inherent variability,
- perform simple probability assessments in order to evaluate the probability of the appropriate performance of engineering activities,
- formulate and solve simple risk based decision problems.

At the end of each chapter, there is a set of small principal exercises which can be used to check and practice the acquired knowledge and skills. The solutions to these are given in the appendix.

Copenhagen, Denmark Michael Havbro Faber

Acknowledgements

The material presented in this book may be appreciated as the outcome of a complex cooking recipe; a great deal of inspiration, some research, a little bit of experience, a lot of help and a ton of time.

Inspiration is what makes us move, in developing the present book, I found inspiration in the many excellent textbooks on applied statistics and probability in engineering. Of course, the classics like Benjamin and Cornell [4] and Ang and Tang [2] play a prominent role here but also the books of Madsen, Krenk and Lind [9], Thoft-Christensen and Baker [13] and Stewart and Melchers [12] should be mentioned—all of them excellent books deserving reading individually.

Research is what changes our perspectives, I have attempted to include a selection of the more recent findings and ideas from research in engineering reliability and risk assessment in the present book. In this connection I would like to thank my research colleagues, not least those from the Joint Committee on Structural Safety.

Experience is what we are, I have spent my professional life, attempting to solve engineering problems and I believe this book has been influenced by that. I would like to thank the real engineers who used their time to teach me.

Help makes us stronger, working alone, I would never have realized this book. I warmly thank especially the members of my research group while I was working at the Swiss Federal Institute of Technology (ETH Zurich): Katharina Fischer, Jochen Köhler, Matthias Schubert, Vasiliki Malioka, Mathias Graf, Markus Deublein, Gerhard Fink, Nishijima Kazuyoshi, Eva Sabiote, Sebastian Thöns, Gianluca De Sanctis, Annett Anders, Gerhard Fink, Harikrishna Narasimhan, Shuoyun Zhang, Jianjun Qin, Daniel Straub, Oliver Kübler, Yahya Bayraktarli and Annette Walzer.

Time is all we have to give, this book took a lot of time to prepare, for me and all other persons involved. Time is precious and for all we know—not recyclable. I dedicate my written works to you who stand behind—my loved children, Emma, Andreas and Ditte—in time you will understand.

Contents

Chapter 1
Engineering Decisions Under Uncertainty

Lecture 1 (Aim of the Present Lecture) The aim of the present lecture is to introduce the problem context of societal decision making and to outline how the concept of risk may provide a means for rational decisions in engineering. Focus is directed to the understanding of the role of the engineer for the development and maintenance of societal functions.
On the basis of the lecture it is expected that the reader will acquire knowledge on the following issues:

- What is sustainability?
- What is the role of engineering in society?
- How can aspects of sustainability be related to life safety and cost optimal decision making?
- What are the main different types of consequences to be considered in risk assessment?
- Why do possible conflicts exist between economy, safety and environment?
- Why is engineering decision making influenced by uncertainties?
- What is the role of probability and consequence in decision making?
- What is the definition of risk?
- Which are the main phases to be considered in life cycle risk assessments in engineering decision making?

1.1 Introduction

During the last two decades, there has been a growing awareness that our world only has limited non-renewable natural resources such as energy and materials but also limited renewable resources such as drinking water, clean air etc. This led the Brundtland Commission in 1987 to the conclusion that *sustainable* development is defined as a development "that meets the needs of the present without compromis-

M.H. Faber, *Statistics and Probability Theory*,
Topics in Safety, Risk, Reliability and Quality 18,
DOI 10.1007/978-94-007-4056-3_1, © Springer Science+Business Media B.V. 2012

ing the ability of future generations to meet their own needs". Sustainable *decision making* is thus presently understood as based on a joint consideration of society, economy and environment. In regard to environmental impacts the immediate implications for the planning, design and operation of civil engineering infrastructures are clear: save energy, save non-renewable resources and find out about re-cycling of building materials, do not pollute the air, water or soil with toxic substances, save or even regain arable land and much more.

For civil engineering infrastructure and facilities in general the financial aspect is also of crucial importance. Civil engineering infrastructure is financed by the public via taxes, public charges or other. In the end, it is the individuals of society who pay and, of course, also enjoy the benefits derived from their existence. However, seen in the light of the conclusions of the Brundtland report (Brundtland [5]), intergenerational equity must be accounted for. Our generation must not leave the burden of maintenance or replacement of too short-lived structures to future generations and it must not use more of the financial resources than those really available. In this sense, civil engineering facilities should be optimal not only from a technological point of view but also from a sustainability point of view.

It is in general a concern how society may maintain and even improve the quality of life. All activities in society should thus aim at improving the life expectancy and increasing the gross domestic product (GDP), resulting in the conclusion that investments into life saving activities must be in balance with the resulting increase in life expectancy. For the present, it is just stated that this problem constitutes a decision problem that can be analyzed using *cost benefit analysis* (see Chap. 7).

At present, approximately 10 to 20% of the GDP of developed countries is being re-invested into life saving activities such as public health, risk reduction and safety. For example, the economic burden of degradation of infrastructure amounted to about 10% of the GDP for the USA in 1997 (see Alsalam et al. [1]). From these numbers it becomes apparent that the issue of safety and well being of the individuals in society as well as the durability of infrastructure facilities has a high importance for the performance of society and the quality of life.

The present book attempts to provide the basic tools for supporting decision making in the context of planning, design and maintenance of civil engineering activities and structures. Engineering facilities such as bridges, power plants, dams and offshore platforms are all intended to benefit, in one way or another the quality of life of the individuals of society. Therefore, whenever such a facility is planned, it is a prerequisite that the benefit of the facility can be proven considering all phases of the life of the facility, i.e. including design, manufacturing, construction, operation and eventually decommissioning. If this is not the case, clearly the facility should not be established.

1.2 Societal Decision Making and Risk

On a societal level a beneficial engineered facility is normally understood as:

- being economically efficient in serving a specific purpose

- fulfilling given requirements with regard to the safety of the personnel directly involved with or indirectly exposed to the facility
- fulfilling given requirements to limit the adverse effects of the facility on the environment.

Based on these requirements, the ultimate task of the engineer is to make decisions or to provide the decision basis for others in order to ensure that engineered facilities are established in such a way as to provide the largest possible benefit.

1.2.1 Example 1.1—Feasibility of Hydraulic Power Plant

Consider as an example the decision problem of exploitation of hydraulic power. A hydraulic power plant project involving the construction of a water reservoir in a mountain valley is planned. The benefit of the hydraulic power plant is for simplicity assumed to be associated only with the monetary income from selling electricity to consumers. The decision problem thus is simplified to compare the costs of establishing, operating and eventually decommissioning the hydraulic power plant with the incomes to be expected during the service life of the plant. In addition, it must be ensured that the safety of the personnel involved in the construction and operation of the plant and the safety of third persons, i.e. the individuals of the society in general, is satisfactorily high.

Different solutions for establishing the power plant may be considered and their efficiency can be measured in terms of the expected income relative to the cost of establishing the power plant. However, a number of factors are important for the evaluation of the income and the costs of establishing the power plant. These are e.g. the period of time when the plant will be operating and produce electricity and the capacity of the power plant. Moreover, the future income from selling electricity will depend on the availability of water, which depends on the future snow and rainfall. Further, the market situation may change and competing energy recourses such as thermal and solar power may cause a reduction of the market price on electricity in general.

In addition, the different possible solutions for establishing the power plant will have different costs and different implications for the safety for persons. Obviously, the more the capacity the power plant will have, the higher the dam and the larger the construction costs will be. At the same time, the potential flooding (consequence of dam failure) will be larger in case of dam failure and more people would be injured or die, see Fig. 1.1.

The safety of the people in a town downstream of the reservoir will also be influenced by the load carrying capacity of the dam structure relative to the pressure loading due to the water level in the reservoir. The strength of the dam structure depends in turn on the material characteristics of the dam structure and the properties of the soil and rock on which it is founded. As these properties are subject to uncertainty of various sources as shall be seen later, the load carrying capacity relative to

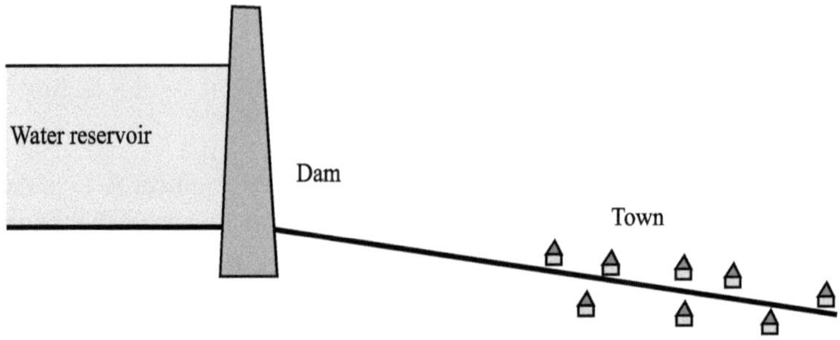

Fig. 1.1 Water reservoir/dam for exploitation of hydraulic power

Table 1.1 Interrelation of benefits, costs and safety for the reservoir

Reservoir water level	Load carrying capacity of dam structure	Income	Costs	Consequence of dam failure	Probability of dam failure
Low	Low	Small	Low	Small	High
	Medium		Medium		Medium
	High		High		Low
Medium	Low	Medium	Low	Medium	High
	Medium		Medium		Medium
	High		High		Low
High	Low	Large	Low	Large	High
	Medium		Medium		Medium
	High		High		Low

the loading may be expressed in terms of the *probability* that the loading will exceed the load carrying capacity, or equivalently, the probability of dam failure.

Finally, the environmental impact of the power plant will depend on the water level in the reservoir: the higher the water level, the more land will be flooded upstream of the dam structure and various habitats for animals and birds will be destroyed. On the other hand, the water reservoir itself will provide a living basis for new species of fish and birds and may provide a range of recreational possibilities for people such as sailing and fishing.

In order to evaluate whether or not the power plant is feasible it is useful to make a list of the various factors influencing the benefit and their effects. As the problem may be recognized to be rather complex, only the interrelation of the water level in the reservoir with the following factors will be considered: the load carrying capacity of the dam structure, the costs of constructing the dam structure and the implications on the safety of the people living in a town down-stream the power plant.

From Table 1.1 which is clearly a simplified summary of the complex interrelations of the various factors influencing the benefit of realizing the power plant, it is seen that the various factors have different influences and that the different attributes

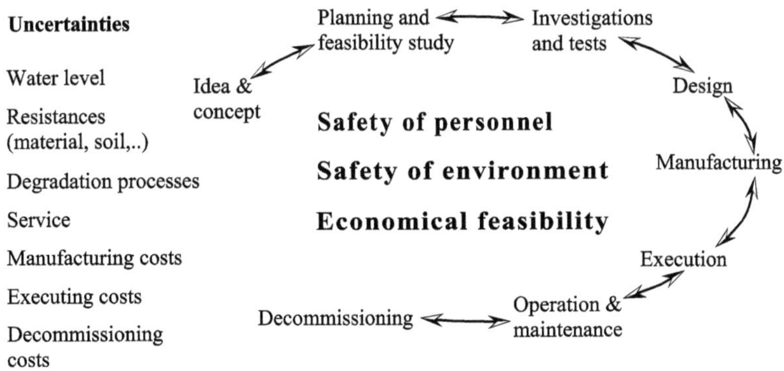

Fig. 1.2 Risk contributions from different service life phases to be considered at the planning stage

such as income, costs and safety are conflicting. In the table it is assumed that the medium load carrying capacity of the dam structure corresponds to a medium probability of dam failure but, of course, other combinations are also possible. Consider the case with a high water level in the reservoir. In this case, the potential income is large but the costs of constructing the dam structure will also be high. Furthermore, the potential *consequences* in case of dam failure will be large as well. Table 1.1 clearly points to the true character of the *decision problem*, namely that the optimal decision depends on the consequences should something go wrong, and moreover, the probability that something goes wrong. The product of these two factors is called the *risk*, a measure that will be considered in much more detail in the chapters to follow.

Furthermore, not only the load carrying capacity of the dam structure is associated with *uncertainty*, but, as indicated previously, also the income expected from the power plant, due to uncertainties in the future market situation. In a similar way, the costs of constructing the power plant are also uncertain as various difficulties encountered during the construction, such as unexpected rock formations, delay in construction works due to problems with material supplies, etc. may lead to additional costs.

When deciding on whether or not to establish the hydraulic power plant it is thus necessary to be able to assess consequences and probabilities—two key factors for the decision problem.

Both consequences and probabilities vary through the life of the power plant and this must be taken into account as well. In the planning phase it is necessary to consider the risk contributions from all subsequent phases of its life-cycle including decommissioning, see Fig. 1.2.

It is important to recognize that different things may go wrong during the different phases of the service life, including events such as mistakes and errors during design and failures and accidents during construction, operation and decommissioning. The potential causes of errors, mistakes, failures and accidents may be numerous, including human errors, failures of structural components, extreme load situations and not least natural hazards. Careful planning during the very first phase of a project is the only way to control the risks associated with such events.

As an illustration, the dam structure must be designed such that the safety of the dam is ensured in all phases of the service life, taking into account yet another factor of uncertainty, namely the future deterioration, but also the quality of workmanship, the degree of quality control implemented during construction and not least the foreseen strategies for the inspection and maintenance of the structures and mechanical equipment during the operation of the power plant. As a final aspect concerning the structures, these should at the end of the service life be in such a condition that the work to be performed during the decommissioning of the power plant can be performed safely for both the persons involved and the environment.

A final fundamental problem arises with regard to the question—how large are the *acceptable risks*?—what is one prepared to invest and/or pay for the purpose of getting a potential benefit? The decision problem of whether or not to establish the hydraulic power plant is thus seen to be a decision problem involving a significant element of uncertainty.

The mathematical basis for the treatment of such decision problems is known as *decision theory*. Important aspects of decision theory are the assessment of consequences and probabilities. In a very simplified manner one can say that risk and reliability analysis in civil engineering is concerned with the problem of decision making subject to uncertainty.

1.3 Definition of Risk

In daily conversation *risk* is a rather common notion used interchangeably with words like *chance*, *likelihood* and *probability* to indicate that people are uncertain about the state of the activity, item or issue under consideration. For example, the risk of getting cancer due to cigarette smoking is discussed, the chance of succeeding to develop a vaccine against the HIV virus in 2020, the likelihood of getting a "Royal Flush" in a Poker game and the probability of a major earthquake occurring in the bay area of San Francisco within the next decade.

Even though it may be understandable from the context of the discussion what is meant by the different words, it is necessary in the context of engineering decision making that those involved are precise in the understanding of risk. Risk is to be understood as the expected consequences associated with a given activity, the activity being e.g. the construction, operation and decommissioning of a power plant.

Considering an activity with only one event with potential consequences C the risk R is the probability that this event will occur P multiplied with the consequences given the event occurs i.e.:

$$R = PC \tag{1.1}$$

If e.g. n events with consequences C_i and occurrence probabilities P_i may result from the activity, the total risk associated with the activity is assessed through the sum of the risks from the individual possible events, i.e.:

$$R = \sum_{i=1}^{n} P_i C_i \tag{1.2}$$

This definition of risk is consistent with the interpretation of risk used e.g. in the insurance industry. Risks may e.g. be given in terms of Euros, Dollars or the number of human fatalities. Even though most risk assessments have some focus on the possible negative consequences of events, the definitions in Eqs. 1.1–1.2 are also valid in the case where benefits are taken into account. In fact, and as will be elaborated in Chap. 7, in this case the definitions in Eqs. 1.1–1.2 are more general and consistent with *expected utility* as basis for decision analysis.

1.4 Self Assessment Questions/Exercises

1. What is meant by the term "sustainable development" and why is it important for engineering decision making?
2. What is normally understood by the term a beneficial engineered facility or activity?
3. How can the risk of an event be defined and how can it be expressed analytically?
4. What is meant by the term "acceptance risks"?
5. Considering an activity with only one event with potential consequences, the associated risk is the probability that this event will occur multiplied with the consequences given the event occurs.
   Which of the following events is associated with the highest risk?

Event	1	2	3
Event probability	0.10	0.01	0.20
Consequences	100 SFr	500 SFr	100 SFr
Risk			

Chapter 2
Basic Probability Theory

Lecture 2 (Aim of the Present Lecture) The aim of the present lecture is to introduce the basics of set and probability theory. Different interpretations of the important concept of probability are provided and it is outlined that the Bayesian probability interpretation facilitates an integration of the other interpretations. The basic axioms of probability theory are given and the important results regarding conditional probabilities and the associated Bayes' rule are outlined.

On the basis of the lecture it is expected that the reader should acquire knowledge and skills with regard to:

- Which are the different interpretations of probability?
- What is a sample space and how can events be illustrated?
- What is an event and what is a complementary event?
- How are intersections and unions of sets defined?
- How can operations involving intersections and unions of events be performed?
- Which are the axioms of probability theory?
- Which are the implications of mutual exclusivity between events?
- What is a conditional probability and how may it be evaluated?
- Which are the implications of independence?
- What is Bayes' rule, and how can it be interpreted?
- How can Bayes' rule be applied for probability updating?

2.1 Introduction

Probability theory forms the basis of the assessment of probabilities of occurrence of uncertain events and thus constitutes a cornerstone in risk and decision analysis. Only when a consistent basis has been established for the treatment of the uncer-

M.H. Faber, *Statistics and Probability Theory*,
Topics in Safety, Risk, Reliability and Quality 18,
DOI 10.1007/978-94-007-4056-3_2, © Springer Science+Business Media B.V. 2012

tainties influencing the probability that events with possible adverse consequences may occur, it is possible to assess the risks associated with a given activity and thus to establish a rational basis for decision making.

The level of uncertainty associated with a considered activity or phenomenon may be expressed by means of purely qualitative statements such as "the chance is good" or "the likelihood is low" but may also be quantified in terms of numbers or percentages. However, the different words all carry the meaning of probability. In the following section this notion, and especially the theoretical framework for its quantification, will be investigated in more detail.

2.2 Definition of Probability

The purpose of the theory of probability is to enable the quantitative assessment of probabilities but the real meaning and interpretation of probabilities and probabilistic calculations is not a part of the theory. Consequently, two people may have completely different interpretations of the probability concept, but still use the same calculus. In the following sections three different interpretations of probability are introduced and discussed based on simple cases. A formal presentation of the axioms of probability theory is provided in Sect. 2.4.

2.2.1 Frequentistic Definition

The *frequentistic definition* of probability is the typical interpretation of probability by the *experimentalist*. In this interpretation the probability $P(A)$ is simply the *relative frequency* of occurrence of the *event* A as observed in an experiment with n trials, i.e. the probability of an event A is defined as the number of times that the event A occurs divided by the number of experiments that are carried out:

$$P(A) = \lim \frac{N_A}{n_{\text{exp}}} \quad \text{for } n_{\text{exp}} \to \infty \tag{2.1}$$

N_A = number of experiments where A occurred, n_{exp} = total number of experiments.

If a frequentist is asked what is the probability of achieving a "head" when flipping a coin, she/he would principally not know what to answer until she/he would have performed a large number of experiments. If say after 1000 experiments (flips with the coin) it is observed that "head" has occurred 563 times, the answer would be that the probability of achieving a "head" is 0.563. However, as the number of experiments is increased the probability would converge towards 0.5. In the mind of a frequentist, probability is a characteristic of nature.

2.2.2 Classical Definition

The *classical probability* definition originates from the days when probability calculus was founded by Pascal and Fermat.[1] The inspiration for this theory can be found in the games of cards and dice. The classical definition of the probability of the event A can be formulated as:

$$P(A) = \frac{n_A}{n_{\text{tot}}} \tag{2.2}$$

n_A = number of equally likely ways by which an experiment may lead to A, n_{tot} = total number of equally likely ways in the experiment.

According to the classical definition of probability, the probability of achieving a "head" when flipping a coin would be 0.5 as there is only one possible way to achieve a "head" and there are two equally likely outcomes of the experiment.

In fact there is no real contradiction to the frequentistic definition, but the following differences may be observed:

- The experiment does not need to be carried out as the answer is known in advance.
- The classical theory gives no solution unless all equally possible ways can be derived analytically.

2.2.3 Bayesian Definition

In the *Bayesian interpretation* the probability $P(A)$ of the event A is formulated as a *degree of belief* that A will occur:

$$P(A) = \text{degree of belief that } A \text{ will occur} \tag{2.3}$$

Coming back to the coin-flipping problem the Bayesian would argue that there are two possibilities, and as she has no preferences as to "head" or "tail" she would judge the probability of achieving a "head" to be 0.5.

The degree of belief is a reflection of the state of mind of the individual person in terms of experience, expertise and preferences. In this respect the Bayesian interpretation of probability is *subjective* or more precise—person-dependent. This opens up the possibility that two different persons may assign different probabilities to a given event and thereby contradicts the frequentist interpretation that probabilities are a characteristic of nature.

The Bayesian statistical interpretation of probability includes the frequentistic and the classical interpretation in the sense that the subjectively assigned probabilities may be based on experience from previous experiments (frequentistic) as well as considerations of e.g. symmetry (classical).

---

[1]Blaise Pascal, philosopher and mathematician, 1623–1662; Pierre de Fermat, jurist and mathematician, 1601–1665. Definition (2.2) is often named after Pierre-Simon de Laplace, mathematician, 1749–1827.

The degree of belief is also referred to as a *prior belief* or *prior probability*, i.e. the belief, which may be assigned prior to obtaining any further knowledge. It is interesting to note that Immanuel Kant[2] developed the philosophical basis for the treatment of subjectivity at the same time as Thomas Bayes[3] developed the mathematical framework later known as *Bayesian statistics*.

Modern structural reliability and risk analysis is based on the Bayesian interpretation of probability. However, the degree of freedom in the assignment of probabilities is in reality not as large as indicated in the above. In a formal Bayesian framework the subjective element should be formulated before the relevant data are observed. Arguments of objective symmetrical reasoning and physical constraints, of course, should be taken into account.

2.2.4 Practical Implications of the Different Interpretations of Probability

In some cases probabilities may adequately be assessed by means of *frequentistic information*. This is e.g. the case when the probability of failure of mass produced components are considered, such as pumps, light bulbs and valves. However, in order to utilize reported failures for the assessment of probability of failure for such components it is a prerequisite that the components are in principle identical, that they have been subject to the same operational and/or loading conditions and that the failures can be assumed to be independent.

In other cases when the considered components are e.g. bridges, high-rise buildings, ship structures or unique configurations of pipelines and pressure vessels, these conditions are not fulfilled. This is because the number of identical structures may be very small (or even just one) and the conditions in terms of operational and loading conditions are normally significantly different from structure to structure. In such situations, the Bayesian interpretation of probability is far more appropriate.

The basic idea behind Bayesian statistics is that lack of knowledge should be treated by probabilistic reasoning, similarly to other types of uncertainty. In reality, decisions have to be made despite the lack of knowledge and probabilistic tools are a great help in this process.

2.3 Sample Space and Events

Considering e.g. the compressive strength of concrete, this material characteristic may be estimated by performing laboratory experiments on standardized test specimens (cylinders or cubes). The test results will, however, be different from one

[2] Immanuel Kant, philosopher, 1724–1804.

[3] Thomas Bayes, reverent and mathematician, 1702–1761.

another and the concrete compressive strength shall thus be assumed to be an uncertain quantity or a random quantity. The set of all possible outcomes of the concrete compressive strength experiments is called the *sample space* (denoted Ω) for the random quantity—the concrete compressive strength. In this example the sample space is the open interval $\Omega =]0; \infty[$, i.e. the set of all positive real numbers. In this case, the sample space is furthermore continuous but in other cases (e.g. when considering the outcome of throwing a dice) the sample space can also be discrete and countable.

An *event* is defined as a subset of a sample space and thus a set of sample points. If the subset is empty (i.e. contains no sample points) it is said to be impossible. An event is said to be certain if it contains all sample points in the sample space (i.e. the event is identical to the sample space).

Consider the events E_1 and E_2 shown in Fig. 2.1. The subset of sample points belonging to the event E_1 or the event E_2 is denoted as the *union* of the events E_1 and E_2 written as $E_1 \cup E_2$.

The subset of sample points belonging to E_1 and E_2 is called the *intersection* of E_1 and E_2 and is written as $E_1 \cap E_2$. The intersection of these two events is illustrated in the right portion of Fig. 2.1.

The two events are said to be *mutually exclusive* if they are disjoint (i.e. if they have no common sample points). In this case, the intersection of E_1 and E_2 is empty (i.e. $E_1 \cap E_2 = \emptyset$), where $\emptyset$ denotes the empty set.

Consider the event E in the sample space Ω. The event containing all sample points in Ω, which are not included in E is called the complementary event to E and denoted $\bar{E}$. It then follows directly that $E \cup \bar{E} = \Omega$ and that $E \cap \bar{E} = \emptyset$.

It can be shown that the intersection and union operations obey the following *commutative, associative and distributive laws*:

$$
\begin{aligned}
E_1 \cap E_2 &= E_2 \cap E_1 \\
E_1 \cap (E_2 \cap E_3) &= (E_1 \cap E_2) \cap E_3 \\
E_1 \cup (E_2 \cup E_3) &= (E_1 \cup E_2) \cup E_3 \\
E_1 \cap (E_2 \cup E_3) &= (E_1 \cap E_2) \cup (E_1 \cap E_3) \\
E_1 \cup (E_2 \cap E_3) &= (E_1 \cup E_2) \cap (E_1 \cup E_3)
\end{aligned}
\tag{2.4}
$$

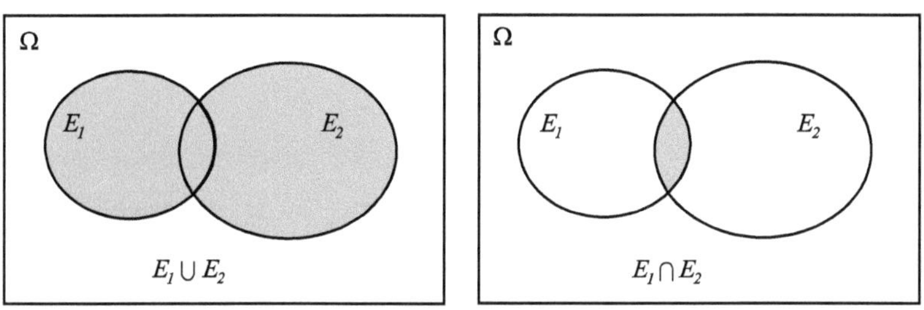

Fig. 2.1 Venn diagrams illustrating the union of events (*left*) and the intersection of events (*right*)

From which the following laws (called *De Morgan's laws*) may be derived:

$$\begin{aligned} E_1 \cap E_2 &= \overline{\bar{E}_1 \cup \bar{E}_2} \\ E_1 \cup E_2 &= \overline{\bar{E}_1 \cap \bar{E}_2} \end{aligned} \tag{2.5}$$

2.4 The Three Axioms of Probability Theory

Probability theory is based on the following three axioms:

Axiom 1

$$0 \le P(E) \le 1 \tag{2.6}$$

for any given event $E \subset \Omega$ where P is the probability measure.

Axiom 2

$$P(\Omega) = 1 \tag{2.7}$$

where Ω is the sample space.

Axiom 3 *Given that $E_1, E_2, \ldots$ is a sequence of mutually exclusive events (i.e. $E_1 \cap E_2 = \emptyset$ etc.) then:*

$$P\left(\bigcup_{i \ge 1} E_i\right) = \sum_{i \ge 1} P(E_i) \tag{2.8}$$

These three *axioms of probability theory* form the sole basis of the theory of probability.

2.5 Conditional Probability and Bayes' Rule

Conditional probabilities are of special interest in risk and reliability analysis as they form the basis of the updating of probability estimates based on new information, knowledge and evidence.

The conditional probability of the event E_1 given that the event E_2 has occurred is written as:

$$P(E_1|E_2) = \frac{P(E_1 \cap E_2)}{P(E_2)} \tag{2.9}$$

It is seen that the conditional probability is not defined if the conditioning event is the empty set, i.e. when $P(E_2) = 0$.

The event E_1 is said to be probabilistically independent of the event E_2 if:

$$P(E_1|E_2) = P(E_1) \tag{2.10}$$

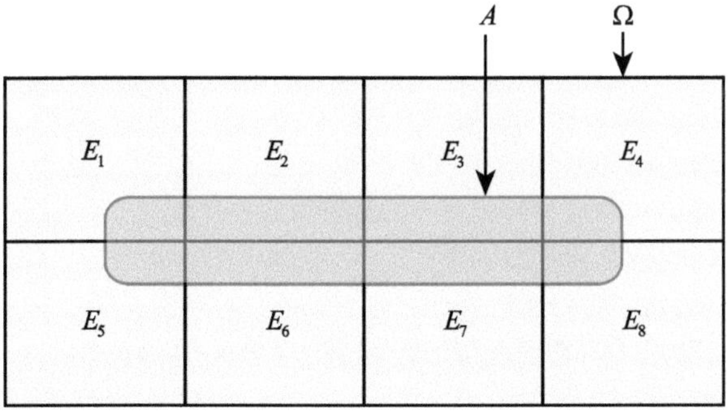

Fig. 2.2 Illustration of the rule of Bayes

implying that the occurrence of the event E_2 does not affect the probability of E_1.

From Eq. 2.9 the probability of the event $E_1 \cap E_2$ may be given as:

$$P(E_1 \cap E_2) = P(E_1|E_2)P(E_2) \tag{2.11}$$

and it follows immediately that if the events E_1 and E_2 are independent, then:

$$P(E_1 \cap E_2) = P(E_1)P(E_2) \tag{2.12}$$

Based on the above findings, the important *Bayes' rule* can be derived.

Consider the sample space Ω divided into n mutually exclusive events E_1, E_2, ..., E_n (see also Fig. 2.2, where the case of $n = 8$ is considered).

Furthermore, let the event A be an event in the sample space Ω. Then the probability of the event A, i.e. $P(A)$, can be written as:

$$\begin{aligned} P(A) &= P(A \cap E_1) + P(A \cap E_2) + \cdots + P(A \cap E_n) \\ &= P(A|E_1)P(E_1) + P(A|E_2)P(E_2) + \cdots + P(A|E_n)P(E_n) \\ &= \sum_{i=1}^{n} P(A|E_i)P(E_i) \end{aligned} \tag{2.13}$$

this is also referred to as the *total probability theorem.*

From Eq. 2.9 there is $P(A|E_i)P(E_i) = P(E_i|A)P(A)$ implying that:

$$P(E_i|A) = \frac{P(A|E_i)P(E_i)}{P(A)} \tag{2.14}$$

Now by inserting Eq. 2.14 into Eq. 2.13, the *Bayes' rule* results:

$$P(E_i|A) = \frac{P(A|E_i)P(E_i)}{\sum_{j=1}^{n} P(A|E_j)P(E_j)} \tag{2.15}$$

In Eq. 2.14 $P(E_i|A)$ is called the *posterior probability* of E_i, the conditional term $P(A|E_i)$ is often referred to as the *likelihood* (i.e. the probability of observing a certain state given the true state). The term $P(E_i)$ is the *prior probability* of the event E_i (i.e. prior to the knowledge about the event A).

As mentioned previously, the Bayes' rule is extremely important, and in order to facilitate its appreciation a few illustrative applications will be given in the following section.

2.5.1 Example 2.1—Using Bayes' Rule for Concrete Assessment

A reinforced concrete beam is considered. From experience it is known that the probability that corrosion of the reinforcement has initiated (the event CI) is $P(CI) = 0.01$. However, in order to know the condition more precisely an inspection method (non-destructive) has been developed.

The quality of the inspection method may be characterized by the probability that the inspection method will indicate initiated corrosion given that corrosion has initiated $P(I|CI)$ (the probability of detection or equivalently the likelihood of an indication I given corrosion initiation CI) and the probability that the inspection method will indicate initiated corrosion given that no corrosion has initiated $P(I|\overline{CI})$ (the probability of erroneous findings or the likelihood of an indication given no corrosion initiation).

For the inspection method at hand the following characteristics have been established:

$$P(I|CI) = 0.8$$
$$P(I|\overline{CI}) = 0.1$$

An inspection of the concrete beam is conducted with the result that the inspection method indicates that corrosion has initiated. Based on the findings from the inspection, what is the probability that corrosion of the reinforcement has initiated?

The answer is readily found by application of Bayes' rule:

$$P(CI|I) = \frac{P(I|CI)P(CI)}{P(I|CI)P(CI) + P(I|\overline{CI})P(\overline{CI})} = \frac{P(I \cap CI)}{P(I)} \tag{2.16}$$

With $P(I)$, the probability of obtaining an indication of corrosion at the inspection:

$$P(I) = P(I|CI)P(CI) + P(I|\overline{CI})P(\overline{CI}) = 0.8 \cdot 0.01 + 0.1 \cdot (1 - 0.01) = 0.107$$

and $P(I \cap CI)$ the probability of receiving an indication of initiated corrosion and at the same time to have initiated corrosion:

$$P(I \cap CI) = P(I|CI)P(CI) = 0.8 \cdot 0.01 = 0.008$$

Thus, the probability that corrosion of the reinforcement has initiated given an indication of initiated corrosion by the inspection method is:

$$P(CI|I) = \frac{0.008}{0.107} = 0.075$$

The probability of initiated corrosion, given an indication of initiated corrosion, is surprisingly low. This is due to the high probability of an erroneous indication of initiated corrosion by the inspection method relative to the small probability of initiated corrosion (i.e. the inspection method is not sufficiently accurate for the considered application).

2.5.2 Example 2.2—Using Bayes' Rule for Bridge Upgrading

An old reinforced concrete bridge is reassessed in connection with an upgrading of the allowable traffic (see also Schneider [11]). The concrete compressive strength class is unknown but concrete cylinder samples may be taken from the bridge and tested in the laboratory.

The following classification of the concrete is assumed:

$$B_1 : 0 \leq \sigma_c < 30$$

$$B_2 : 30 \leq \sigma_c < 40$$

$$B_3 : 40 \leq \sigma_c$$

where σ_c is the compressive strength of concrete.

Even though the concrete class is unknown, experience with similar bridges suggests that the probability of the concrete of the bridge belonging to class B_1, B_2 and B_3 is 0.65, 0.24 and 0.11, respectively. This information comprises the prior information—prior to any experiment result.

The test method is not perfect in the sense that even though the test indicates a value of the concrete compressive strength belonging to a certain class, there is a certain probability that the concrete belongs to another class. The likelihoods for the considered test method are given in Table 2.1.

It is assumed that one test is performed and it is found that the concrete compressive strength is equal to 36.2 MPa, i.e. in the interval of class B2.

Using Bayes' rule, the probability that the concrete belongs to one of the different classes may now be updated. The posterior probability that the concrete belongs to class B_2 is given by:

$$P(B_2|I = B_2) = \frac{0.61 \cdot 0.24}{0.61 \cdot 0.24 + 0.28 \cdot 0.65 + 0.32 \cdot 0.11} = 0.40$$

The posterior probabilities for the other classes may be calculated in a similar manner, the results are given in Table 2.1.

Table 2.1 Summary of prior probabilities, likelihoods of experiment outcomes and posterior probabilities given one test result in the interval of class B_2

Concrete class	Prior probability	Likelihood $P(I\|B_i)$			Posterior probability
		$I = B_1$	$I = B_2$	$I = B_3$	
B_1	0.65	0.71	0.28	0.01	0.50
B_2	0.24	0.18	0.61	0.21	0.40
B_3	0.11	0.02	0.32	0.66	0.10

2.6 Self Assessment Questions/Exercises

1. A person is asked what is the probability for achieving a "head" when flipping a coin. The person after 1000 experiments (flips with the coin) observes that "head" has occurred 333 times and hence answers that the probability for "head" is 0.333. On which interpretation of probability is this estimation based on?
2. How can the conditional probability of an event E_1, given that the event E_2 has occurred, be written?
3. In probability theory the probability, $P(A)$, of an event A can take any value within the following boundaries:
   $0 \leq P(A) \leq 1$ ☐
   $-1 \leq P(A) \leq 1$ ☐
   $-\infty < P(A) < \infty$ ☐
4. If the intersection of two events, A and B corresponds to the empty set $\emptyset$, i.e. $A \cap B = \emptyset$, the two events are:
   Mutually exclusive. ☐
   Independent. ☐
   Empty events. ☐
5. Which the following expressions is(are) correct?
   The probability of the union of two events A and B is equal to the sum of the probability of event A and the probability of event B, given that the two events are mutually exclusive. ☐
   The probability of the union of two events A and B is equal to the probability of the sum of the two events A and event B, given that the two events are mutually exclusive. ☐
   The probability of the intersection of two events A and B is equal to the product of the probability of event A and the probability of event B, given that the two events are mutually exclusive. ☐
   The probability of the intersection of two events A and B is equal to the product of the probability of event A and the probability of event B, given that the two events are independent. ☐
6. The probability of the intersection of two mutually exclusive events is equal to:
   The product of the probabilities of the individual events. ☐
   The sum of the probabilities of the individual events. ☐

The difference between the probabilities of the individual events. ☐
One (1). ☐
Zero (0). ☐

7. Which of the following statements is correct?
An event A is defined as a subset of a sample space Ω. ☐
A sample space Ω is defined as a subset of an event A. ☐
8. The probability of the union of two not mutually exclusive events A and B is given as: $P(A \cup B) = P(A) + P(B) - P(A \cap B)$. It is provided that the probability of event A is equal to 0.1, the probability of event B is 0.1 and the probability of event B given event A, i.e. $P(B|A)$ is 0.8. Which result is correct?
$P(A \cup B) = -0.6$ ☐
$P(A \cup B) = 0.12$ ☐
$P(A \cup B) = 0.04$ ☐
9. For an event A in the sample space Ω, event $\bar{A}$ represents the complementary event of event A. Which one(s) of the following expressions are correct?
$A \cup \bar{A} = \Omega$ ☐
$A \cap \bar{A} = \Omega$ ☐
$A \cup \bar{A} = \emptyset$ ☐
10. The commutative, associative and distributive laws describe how to:
Operate with intersections of sets. ☐
Operate with unions of sets. ☐
None of the above. ☐
11. Research in ETH is often funded by the Swiss National Foundation of research (SNF). The normal procedure is that a Professor submits a proposal for a new project. Experts working for SNF read the proposal and they may come to one of the following decisions:
D_1: the proposal is accepted and the project will be funded.
D_2: the proposal should be revised by the Professor and resubmitted to SNF.
D_3: the proposal is not accepted and hence no funding is provided.
Professor Muster works at ETH. During the past few years he has submitted many proposals to SNF. Based on experience, over many years, Professor Muster in general assesses that when he submits a proposal the probabilities associated with the possible final decisions of SNF are as follows:

$$P(D_1) = 0.45, \qquad P(D_2) = 0.35, \qquad P(D_3) = 0.2$$

By coincidence, just at the time when Professor Muster considers to submit a new proposal to SNF, he meets Dr. Beispiel. Dr. Beispiel used to work at SNF as one of the experts who review proposals and make the final decisions. Professor Muster kindly asks Dr. Beispiel to have a look at the new proposal before submitting it to SNF with the purpose of assessing the probabilities that the proposal would be accepted as it is. Of course Dr. Beispiel cannot say with certainty what will be the final SNF decision. However, his assessment can be considered as an indication, I_j, of the final decision of SNF. Based on experience from previous assessments and final decisions the conditional probabilities, $P(I_j|D_i)$, of the indications I_j of Dr. Beispiel given the final decisions D_i of SNF are as summarized in the following table.

SNF final decision D_i	Dr. Beispiel's indicative assessment I_j		
	I_1	I_2	I_3
D_1	$P(I_1 \mid D_1) = 0.86$	$P(I_2 \mid D_1) = 0.1$	
D_2	$P(I_1 \mid D_2) = 0.2$		$P(I_3 \mid D_2) = 0.06$
D_3		$P(I_2 \mid D_3) = 0.1$	$P(I_3 \mid D_3) = 0.9$

a. Complete the above table.
b. Having read the new proposal Dr. Beispiel explains to Professor Muster that if he would still have been working with SNF he would have asked for revisions and resubmission. Based on this new information—what is the probability that the final decision of SNF is the same as the assessment of Dr. Beispiel?

Chapter 3
Descriptive Statistics

Lecture 3 (Aim of the Present Lecture) The aim of the present lecture is to introduce descriptive statistics in terms of numerical summaries and graphical representations. It is outlined how data can be represented in a standardized manner numerically as well as in the form of different graphs.
On the basis of the lecture it is expected that the reader should acquire knowledge and skills with regard to:

- What is the purpose of descriptive statistics?
- In what principally different ways can data be assessed and communicated?
- What are the assumptions underlying descriptive statistics?
- Which are the different "central numerical measures" and what do they describe?
- What is a measure of dispersion and which such measures are available?
- What does peakedness and skewness refer to?
- What is the significance of correlation and how may it be calculated?
- Which are typical graphical representations of data sets?
- What is the difference between a sample histogram and a frequency distribution?
- What information is contained in a Quantile-Quantile plot?
- What are the main components of a Tukey box plot?
- In what way can numerical summaries be related to graphical representations?

3.1 Introduction

In order to assess the knowledge of a given quantity of interest, one of the first steps is to investigate the data available, such as *observations* and *test results*. For this purpose, descriptive statistics is useful. *Descriptive statistics* do not assume anything

M.H. Faber, *Statistics and Probability Theory*,
Topics in Safety, Risk, Reliability and Quality 18,
DOI 10.1007/978-94-007-4056-3_3,

in terms of the degree or nature of the randomness underlying the data analyzed but are merely a convenient tool to reduce the data to a manageable form suitable for further analysis, as well as for communication of the data in a standardized format to other professionals.

In the following section *numerical summaries* will first be introduced. These can be considered to be numerical characteristics of the observed data containing important information about the data and the nature of uncertainty associated with these. These are also referred to as *sample characteristics*. Thereafter, *graphical representations* are introduced as means of visual characterization and as a useful tool for data analysis. Descriptive statistics play an important role in engineering risk analysis as a standardized basis for assessing and documenting data obtained for the purpose of understanding and representing uncertainties in risk assessment.

3.2 Numerical Summaries

3.2.1 Central Measures

One of the most useful numerical summaries is the *sample mean*. If the data set is collected in the vector $\hat{\mathbf{x}} = (\hat{x}_1, \hat{x}_2, \ldots, \hat{x}_n)^T$ the sample mean $\bar{x}$ is simply given as:

$$\bar{x} = \frac{1}{n}\sum_{i=1}^{n} \hat{x}_i \qquad (3.1)$$

The sample mean may be interpreted as a central value of the data set. If, on the basis of the data set, one should give only one value characterizing the data, one would normally use the sample mean.

Another central measure is the mode of the data set i.e. the most frequently occurring value in the data set. When data samples are real values, the *mode* in general cannot be assessed numerically, but may be assessed from graphical representations of the data as will be illustrated in Sect. 3.3.

It is often convenient to work with an ordered data set which is readily established by rearranging the original data set $\hat{\mathbf{x}} = (\hat{x}_1, \hat{x}_2, \ldots, \hat{x}_n)^T$ such that the data are arranged in increasing order as $\hat{x}_1^o \le \hat{x}_2^o \le \cdots \le \hat{x}_i^o \le \cdots \le \hat{x}_{n-1}^o \le \hat{x}_n^o$. The ith value of an ordered data set is denoted by $\hat{x}_i^o$.

The *median* of the data set is defined as the middle value in the ordered list of data if n is odd. If n is even the median is taken as the average value of the two middle values (see also the examples below).

3.2.2 Example 3.1—Concrete Compressive Strength Data

Consider the data set given in Table 3.1 corresponding to concrete cube compressive strength measurements. In the table the data are listed both unordered, e.g. in the order they were observed and ordered according to increasing values.

Table 3.1 Concrete cube compressive strength experiment results in MPa

i	Unordered $\hat{x}_i$	Ordered $\hat{x}_i^o$
1	35.8	24.4
2	39.2	27.6
3	34.6	27.8
4	27.6	27.9
5	37.1	28.5
6	33.3	30.1
7	32.8	30.3
8	34.1	31.7
9	27.9	32.2
10	24.4	32.8
11	27.8	33.3
12	33.5	33.5
13	35.9	34.1
14	39.7	34.6
15	28.5	35.8
16	30.3	35.9
17	31.7	36.8
18	32.2	37.1
19	36.8	39.2
20	30.1	39.7

The sample mean for the data set is readily evaluated using Eq. 3.1 and found to be equal to 32.67 MPa. All the observed values are different and therefore the mode cannot be determined without dividing the observations into intervals as will be shown in Sect. 3.3. However, the median is readily determined as being equal to 33.05 MPa.

3.2.3 Example 3.2—Traffic Flow Data

Consider the data shown in Table 3.2. The data correspond to the daily traffic flow in both directions through the Gotthard tunnel for the month of January 1997 obtained within a project carried out by the Swiss Federal Highways Office (ASTRA).

For this data set the sample mean values of the traffic flow in direction 1 and direction 2 may be calculated from either the unordered or ordered data sets to be equal to 4697.39 and 5660.77 respectively. The corresponding median values can be read from the ordered data sets as 4419 and 5100 respectively (there are in total 31 observations in the data sets, so the median corresponds to observation 16).

Table 3.2 Daily traffic flow through the Gotthard tunnel, January 1997

i	Direction 1			Direction 2		
	date	unordered $\hat{x}_i$	ordered $\hat{x}_i^o$	date	unordered $\hat{x}_i$	ordered $\hat{x}_i^o$
1	01.01	3087	3087	01.01	3677	3677
2	02.01	4664	3578	02.01	7357	4453
3	03.01	4164	3710	03.01	9323	4480
4	04.01	3710	3737	04.01	11748	4560
5	05.01	4029	3906	05.01	10256	4635
6	06.01	4323	4029	06.01	4453	4648
7	07.01	4041	4041	07.01	4815	4672
8	08.01	3737	4085	08.01	4757	4757
9	09.01	4103	4103	09.01	4672	4791
10	10.01	5457	4164	10.01	5401	4815
11	11.01	4563	4323	11.01	5688	4880
12	12.01	3906	4359	12.01	6308	4928
13	13.01	4419	4366	13.01	4946	4946
14	14.01	4359	4368	14.01	4635	5005
15	15.01	4667	4371	15.01	5100	5013
16	16.01	5098	4419	16.01	4791	5100
17	17.01	6551	4563	17.01	5235	5220
18	18.01	4371	4588	18.01	4560	5235
19	19.01	3578	4664	19.01	5729	5281
20	20.01	4366	4667	20.01	5005	5318
21	21.01	4368	4727	21.01	4480	5398
22	22.01	4588	4739	22.01	4880	5401
23	23.01	5001	4741	23.01	4928	5679
24	24.01	7118	5001	24.01	5398	5688
25	25.01	4727	5098	25.01	4648	5729
26	26.01	4085	5193	26.01	6183	6183
27	27.01	4741	5457	27.01	5220	6308
28	28.01	4739	5892	28.01	5013	7357
29	29.01	5193	6551	29.01	5281	9323
30	30.01	5892	7118	30.01	5318	10256
31	30.01	7974	7974	31.01	5679	11748

3.2.4 Dispersion Measures

The variability or the *dispersion* of the data set around the sample mean is also an important characteristic of the data set. The dispersion may be characterized by the *sample variance* s^2 given by:

$$s^2 = \frac{1}{n} \sum_{i=1}^{n} (\hat{x}_i - \bar{x})^2 \tag{3.2}$$

The *sample standard deviation* s is defined as the square root of the sample variance. From Eq. 3.2 it is seen that the sample standard deviation s is assessed in terms of the variability of the observations around the sample mean value $\bar{x}$.

Thus, the sample variance is the mean of the squared deviations from the sample mean and is in this way analogous to the moment of inertia as used in structural engineering.

As a means of comparison of the dispersions of different data sets, the dimensionless *sample coefficient of variation* ν is convenient. The sample coefficient of variation ν is defined as the ratio of the sample standard deviation to the sample mean, i.e. given by:

$$\nu = \frac{s}{\bar{x}} \tag{3.3}$$

The *sample variance* for the concrete cube compressive strengths of Table 3.1 may be evaluated using Eq. 3.2 and is found to be 16.36 MPa^2. The sample standard deviation is thus 4.04 MPa. For the considered concrete cube compressive strength data the sample coefficient of variation is equal to 0.12. In the same manner the sample coefficient of variation for the traffic flow data in Table 3.2 is equal to 0.21 and 0.30 for direction 1 and direction 2 respectively. It is seen that the coefficient of variation for direction 2 is higher than for direction 1. That indicates that the data observed in direction 2 are more dispersed than in direction 1.

3.2.5 Other Measures

Whereas the sample mean, mode and median are central measures of a data set and the sample variance is a measure of the dispersion around the sample mean, it is also useful to have some characteristic indicating the degree of symmetry of the data set. To this end, the sample coefficient of *skewness*, which is a simple logical extension of the sample variance is suitable. The sample coefficient of skewness η is defined as:

$$\eta = \frac{1}{n} \frac{\sum_{i=1}^{n} (\hat{x}_i - \bar{x})^3}{s^3} \tag{3.4}$$

This coefficient is positive if the mode of the data set is less than its mean value (skewed to the right) and negative if the mode is larger than the mean value (skewed to the left). For the concrete cube compressive strengths (Table 3.1) the sample coefficient of skewness is −0.12. For the traffic flow data (Table 3.2) the observations in directions 1 and 2 have a skewness coefficient of 1.54 and 2.25 respectively. The coefficients are positive and that show that both distributions are skewed to the right.

In a similar way the sample coefficient of *kurtosis* κ is defined as:

$$\kappa = \frac{1}{n}\frac{\sum_{i=1}^{n}(\hat{x}_i - \bar{x})^4}{s^4} \tag{3.5}$$

which is a measure of how closely the data are distributed around the mode (peakedness). Typically one would compare the sample coefficient of kurtosis to that of a normal distribution (introduced in Chap. 4), which is equal to 3.0. The kurtosis for the concrete cube compressive strength (Table 3.1) is evaluated as equal to 2.23, i.e. the considered data set is less peaked than the normal distribution. For the traffic flow data (Table 3.2) it is equal to 5.48 and 7.44 for directions 1 and 2 respectively.

3.2.6 Sample Moments and Sample Central Moments

In Eqs. 3.2, 3.4 and 3.5 the sample central moments are used. These are denoted as "central" as they are always taken about mean. The equations show that the sample mean $\bar{x}$ is subtracted from the particular observations $\hat{x}_i$. In general the central sample moments can be defined as:

$$z_j = \frac{1}{n}\sum_{i=1}^{n}(\hat{x}_i - \bar{x})^j$$

In Eqs. 3.4 and 3.5 the 3rd and 4th central sample moment are divided by the quadratic standard deviation to assess the skewness and the kurtosis. The "simple" (non-central) sample moments do not use the sample mean in their definition. Hence, the general definition is:

$$m_j = \frac{1}{n}\sum_{i=1}^{n}\hat{x}_i^j$$

3.2.7 Measures of Correlation

Observations are often made of two characteristics simultaneously as shown in Fig. 3.1 where pairs of data observed simultaneously are plotted jointly along the x-axis and the y-axis (this representation is also called a two-dimensional *scatter diagram* as outlined in Sect. 3.3).

As a characteristic indicating the tendency toward high-high pairings and low-low pairings, i.e. a measure of the *correlation* between the observed data sets, the *sample covariance* s_{XY} is useful, and is defined as:

$$s_{XY} = \frac{1}{n}\sum_{i=1}^{n}(\hat{x}_i - \bar{x})(\hat{y}_i - \bar{y}) \tag{3.6}$$

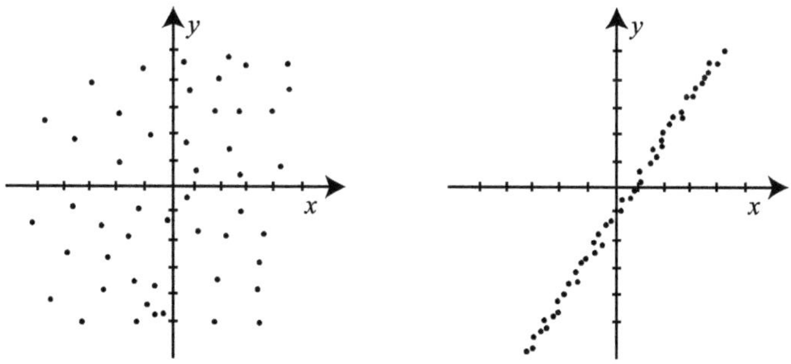

Fig. 3.1 Two examples of paired data sets

The sample covariance has the property that, if there is a tendency in the data set that the values of $\hat{x}_i$ and $\hat{y}_i$ are both higher than $\bar{x}$ and $\bar{y}$ at the same time, and the trend is linear, then most of the terms in the sum will be positive and the sample covariance will be positive. The opposite trend will result in a negative sample covariance. Such behavior is referred to as correlation.

In the scatter diagram to the left in Fig. 3.1 there is only little correlation between the observed data pairs whereas the opposite is evident in the example to the right.

The sample covariance may be normalized with respect to the sample standard deviations of the individual data sets s_X and s_Y and the result is called the sample *correlation coefficient* r_{XY} defined as:

$$r_{XY} = \frac{1}{n} \frac{\sum_{i=1}^{n} (\hat{x}_i - \bar{x})(\hat{y}_i - \bar{y})}{s_X s_Y} \tag{3.7}$$

The sample correlation coefficient has the property that it is limited to the interval $-1 \leq r_{XY} \leq 1$ and the extreme values of the interval are only achieved in case the data pairs are perfectly correlated, implying that the points on the scatter diagram lie on a straight line. For the example shown in Fig. 3.1 there is almost zero correlation at the left hand side and almost full positive correlation at the right hand side.

3.3 Graphical Representations

Graphical representations provide a convenient basis for assessing data as well as communicating these to other persons. There exist a relatively large number of different possible graphical representations of data, of which some are better suited than others depending on the purpose of the representations. Some are better for representing the characteristics of data sets containing observations of one characteristic, e.g. the concrete compressive strength and others are better for representing the characteristics of two or more data sets e.g. the simultaneously observed traffic flows. In the following sub-sections the most frequently applied graphical representations are introduced and discussed with the help of examples.

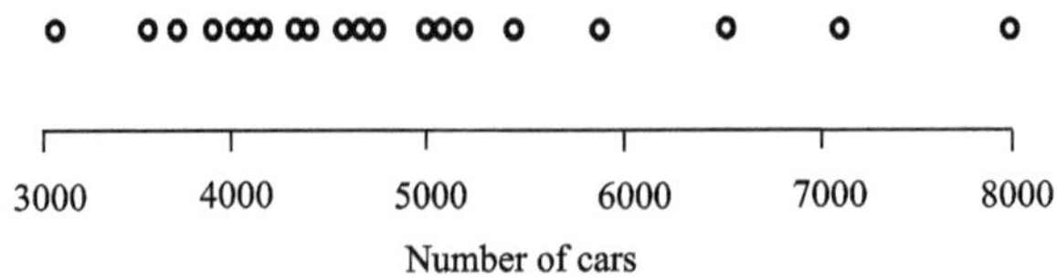

Fig. 3.2 One-dimensional scatter diagram of the traffic flow in the Gotthard tunnel (direction 1)

3.3.1 One-Dimensional Scatter Diagrams

The simplest graphical representation is the *scatter diagram* which provides a means to represent observations contained in one or more data sets. The scatter diagram may be constructed by plotting the observed values of the data set along an axis labeled according to the scale of the observations. In a *one-dimensional scatter diagram* the minimum and maximum values of the data set can be readily observed. Furthermore, as long as the number of data points is not very large, the central value of the observed data may be observed directly from the plot. In the case where a data set contains a large number of data, some of these may be overlapping and this makes it difficult to distinguish the individual observations. In such cases, it may be beneficial to apply another graphical representation such as histograms, as described subsequently.

Consider again the traffic flow data from Table 3.2. For each of the two directions a one-dimensional scatter diagram can be produced by plotting the data along one axis. In Fig. 3.2 the resulting scatter diagram is shown for direction 1. It can be seen from Fig. 3.2 that the lowest value of the data lies close to 3000 while the highest lies close to 8000. Moreover, a high concentration of observations is observed in the range 4000 to 5000, indicating that the central value of this data is in that range.

3.3.2 Histograms

A frequently applied graphical representation of data sets is the *histogram*. Consider again as an example the traffic flow data from Table 3.2 for direction 2. The data are further processed and the observed number of cars is subdivided into intervals, see Table 3.3. For each interval the midpoint is determined and the number of observations within each interval is counted. Thereafter, the *frequencies* of the measurements within each interval are evaluated as the number of observations within one interval divided by the total number of observations. The *cumulative frequencies* are estimated by summing up the frequencies for each interval in increasing order. This is a common way to estimate the cumulative frequencies especially in cases where the exact observations are not known but instead the frequency of observations within an interval is known. In the following, for illustration purposes, the cumulative frequencies are estimated from available observations. However, when observations are readily available the cumulative frequency plot can be replaced by

Table 3.3 Summary of the observed traffic flow in the Gotthard tunnel (direction 2)

Interval (number of cars) $[10^2/\text{day}]$	Interval midpoint (number of cars) $[10^2/\text{day}]$	Number of observations	Frequency [%]	Cumulative frequency
30–40	37.5	1	3.2258	0.0323
40–45	42.5	2	6.4516	0.0968
45–50	47.5	10	32.2581	0.4194
50–55	52.5	9	29.0323	0.7097
55–60	57.5	3	9.6774	0.8065
60–65	62.5	2	6.4516	0.8710
65–70	67.5	0	0.0000	0.8710
70–75	72.5	1	3.2258	0.9032
75–80	77.5	0	0.0000	0.9032
80–85	82.5	0	0.0000	0.9032
85–90	87.5	0	0.0000	0.9032
90–95	92.5	1	3.2258	0.9355
95–100	97.5	0	0.0000	0.9355
100–105	102.5	1	3.2258	0.9677
105–110	107.5	0	0.0000	0.9677
110–115	112.5	0	0.0000	0.9677
115–120	117.5	1	3.2258	1.0000

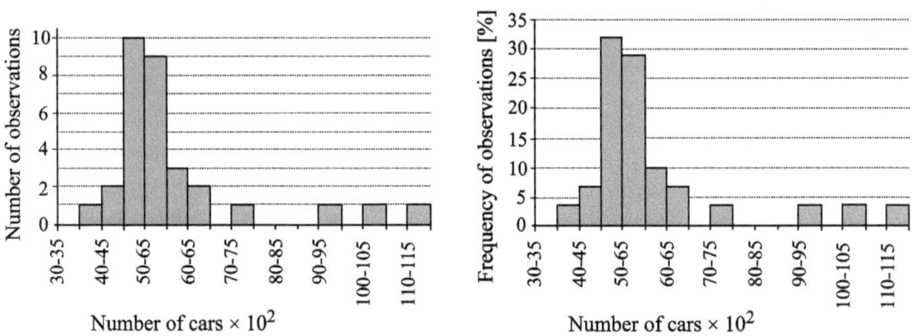

Fig. 3.3 Histogram and frequency distribution representations of the observed traffic flow in the Gotthard tunnel (direction 2)

a plot similar to a quantile plot (see Sect. 3.3.3) but a slightly different representation. Figures 3.3 and 3.4 show the graphical representation of the processed data of Table 3.3.

It has to be noted that a histogram may reduce the information provided by the data examined. The interval width plays an important role for the resolution of the representation of the observations. However, there are no general guidelines con-

Table 3.4 Summary of the observed traffic flow in the Gotthard tunnel (direction 2) using 6 intervals

Interval (number of cars) $[10^2/\text{day}]$	Interval midpoint (number of cars) $[10^2/\text{day}]$	Number of observations	Frequency [%]	Cumulative frequency
35–50	42.5	13	41.9355	0.4194
50–65	57.5	14	45.1613	0.8710
65–80	72.5	1	3.2258	0.9032
80–95	87.5	1	3.2258	0.9355
95–110	102.5	1	3.2258	0.9677
110–125	117.5	1	3.2258	1.0000

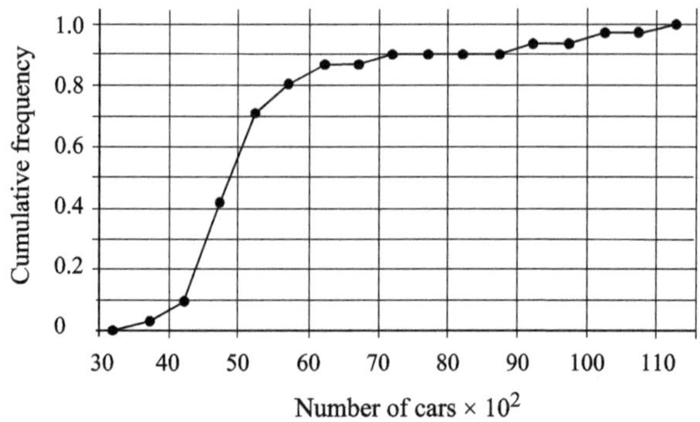

Fig. 3.4 Cumulative frequency plot of the observed traffic flow in the Gotthard tunnel (direction 2)

cerning the choice of the interval width. In most applications, the goal is to identify an interval, which with a sufficient resolution, can represent the observations and this may comprise an iterative process where several different subdivisions are applied and the results are evaluated. In Benjamin and Cornell [4] it is suggested to subdivide the interval between the maximum and minimum value into k intervals where k is given by:

$$k = 1 + 3.3\log_{10}(n) \tag{3.8}$$

where n is the number of data points in the data set.

Using the above formula for the observations in Table 3.2, k equals 5.92. By rounding up, 6 intervals should have been applied for the subdivision of observations while as shown in Table 3.3 the number of intervals used is equal to 17. Figure 3.5 illustrates the frequency distribution of the traffic flow data using the 6 intervals given in Table 3.4.

Compared with the frequency distribution in Fig. 3.5 it can be seen that using a smaller number of intervals the resolution of the graphical representation is significantly reduced.

Table 3.5 Summary of the observed concrete cube compressive strength measurements

Interval (compressive strength) [MPa]	Interval midpoint (compressive strength) [MPa]	Number of observations	Frequency [%]	Cumulative frequency
23–26	24.5	1	5	0.05
26–29	27.5	4	20	0.25
29–32	30.5	3	15	0.40
32–35	33.5	6	30	0.70
35–38	36.5	4	20	0.90
38–41	39.5	2	10	1.00

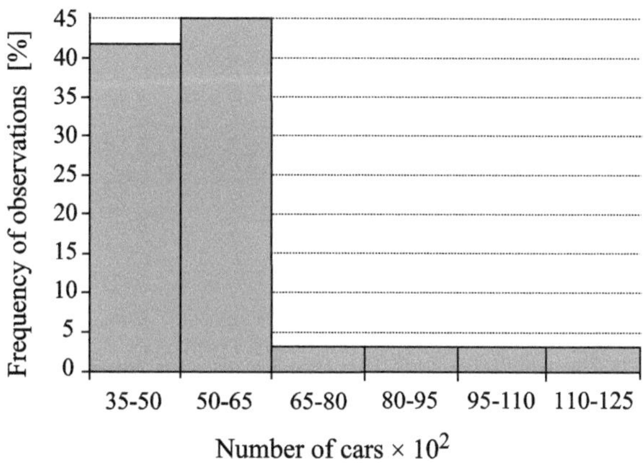

Fig. 3.5 Frequency distribution representation of the observed traffic flow in the Gotthard tunnel (direction 2) using a small number of intervals

For the concrete compressive strength observations the application of Eq. 3.8 seems to work well. Equation 3.8 in this case gives a value of $k = 5.29$ and by rounding up 6 intervals should be used for this data set. The data organized according to the result of Eq. 3.8 are given in Table 3.5.

Figures 3.6 and 3.7 show the graphical representation of the processed data of Table 3.5. It is seen from Fig. 3.6 that the rule implied from Eq. 3.8 works fine and the resulting frequency distribution provides a good resolution of the observations.

3.3.3 *Quantile Plots*

Quantile plots are graphical representations containing information that is similar to the cumulative frequency plots introduced above. A quantile is related to a given percentage, e.g. the 0.65 quantile of a given data set of observations is the observation for which 65% of all observations in the data set have smaller values. The

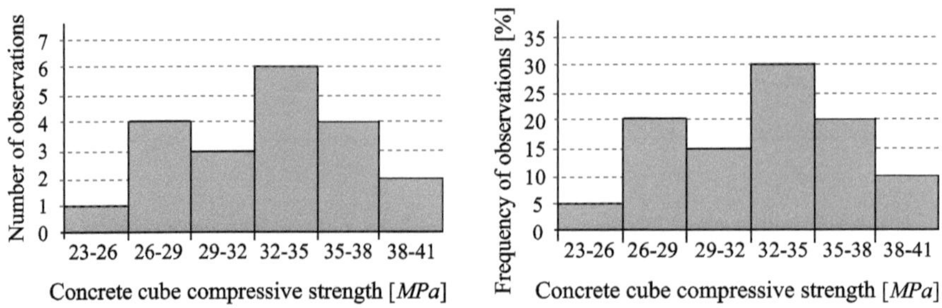

Fig. 3.6 Histogram and frequency distribution representations of the observed concrete cube compressive strength

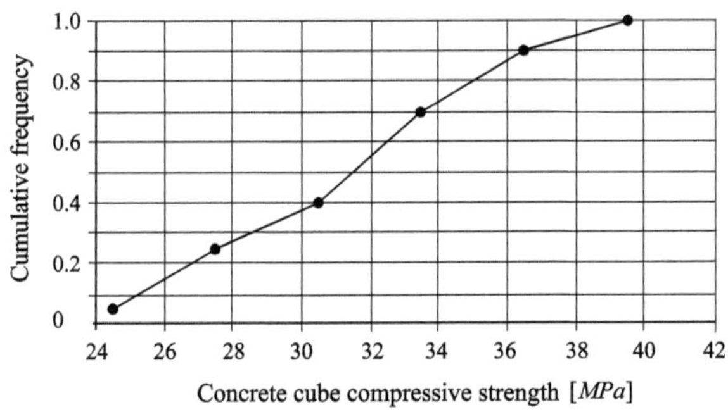

Fig. 3.7 Cumulative frequency plot of the observed concrete cube compressive strength

0.75 quantile is also called the *upper quartile* (see also the Tukey box plots in the next section) while the 0.25 quantile is called the *lower quartile*. The median is thus equal to 0.5 quantile.

In order to construct a quantile plot the observations in the data set are arranged in ascending order. The quantile Q_υ is equal to a given observation $\hat{x}_i^o$ in the ordered data set where the index of the quantile, υ, is given by:

$$\upsilon = \frac{i}{n+1} \tag{3.9}$$

As an example consider the traffic flow data from Table 3.2. In Table 3.6 the data are ordered in ascending order and the corresponding quantile indices are shown. The median (i.e. the 0.5 quantile) has been highlighted.

As mentioned in Sect. 3.3.2, when the observations are known it is preferable to use their quantiles to represent the cumulative distribution instead of the frequency of observations within interval. So for example, in the case of the traffic flow data for direction 2 (Table 3.4) the cumulative distribution can be plotted the data values and the respective quantile indices. Similarly, the cumulative distribution of the concrete cube compressive strength data can be plotted using the respective quantile indices shown in Table 3.7. Figure 3.8 illustrates the two cumulative distribution

Table 3.6 Quantile values of the traffic flow observations in the Gotthard tunnel

i	Direction 1 ordered $\hat{x}_i^o$	Direction 2 ordered $\hat{x}_i^o$	υ
1	3087	3677	0.0313
2	3578	4453	0.0625
3	3710	4480	0.0938
4	3737	4560	0.1250
5	3906	4635	0.1563
6	4029	4648	0.1875
7	4041	4672	0.2188
8	4085	4757	0.2500
9	4103	4791	0.2813
10	4164	4815	0.3125
11	4323	4880	0.3438
12	4359	4928	0.3750
13	4366	4946	0.4063
14	4368	5005	0.4375
15	4371	5013	0.4688
16	4419	5100	0.5000
17	4563	5220	0.5313
18	4588	5235	0.5625
19	4664	5281	0.5938
20	4667	5318	0.6250
21	4727	5398	0.6563
22	4739	5401	0.6875
23	4741	5679	0.7188
24	5001	5688	0.7500
25	5098	5729	0.7813
26	5193	6183	0.8125
27	5457	6308	0.8438
28	5892	7357	0.8750
29	6551	9323	0.9063
30	7118	10256	0.9375
31	7974	11748	0.9688

plots mentioned above. The median of the data set can be directly read from such a representation by finding the value that corresponds to the 0.5 quantile.

The cumulative frequency plots can be distinguished from the quantile plots by the different axes titles. The x-axis of the quantile plot contains the quantile index, which in contrast can be found on the y-axis of the cumulative frequency plot. The latter connects the particular points of cumulative frequency values by drawing steps

Table 3.7 Quantile values of the observed concrete cube compressive strength [MPa]

i	Ordered $\hat{x}_i^o$	υ
1	24.4	0.048
2	27.6	0.095
3	27.8	0.143
4	27.9	0.190
5	28.5	0.238
6	30.1	0.286
7	30.3	0.333
8	31.7	0.381
9	32.2	0.429
10	32.8	0.476
11	33.3	0.524
12	33.5	0.571
13	34.1	0.619
14	34.6	0.667
15	35.8	0.714
16	35.9	0.762
17	36.8	0.810
18	37.1	0.857
19	39.2	0.905
20	39.7	0.952

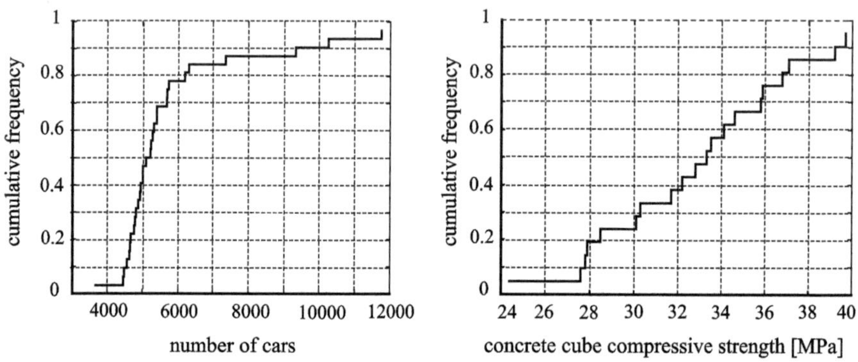

Fig. 3.8 Cumulative distribution plot of the traffic flow data of direction 2 (*left*) and the concrete cube compressive strength (*right*)

in between (Fig. 3.8). The quantile plot simply illustrates the values by unconnected points (Fig. 3.9).

In Fig. 3.9 the quantile plots for the traffic flow data for both directions are illustrated. In order to facilitate the comparison of the data these have been plotted

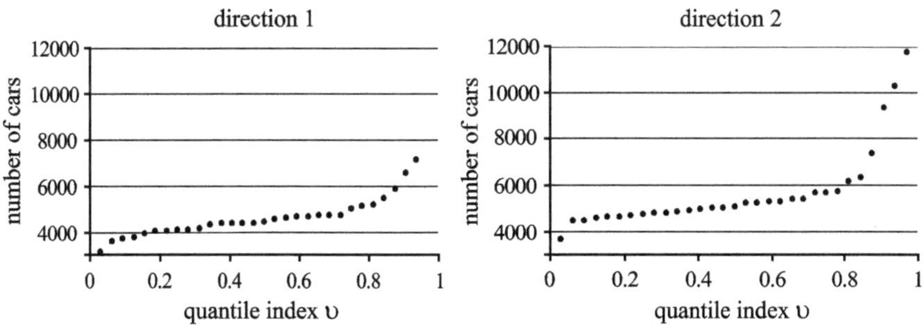

Fig. 3.9 Quantile plots of the observed traffic flow in the Gotthard tunnel

on the same scale. It can be seen that the quantiles for the data in direction 2 are slightly higher than the corresponding ones for direction 1. For direction 1 the median (the 0.5 quantile) is close to 4500 (the real value read from Table 3.6 is 4419). The corresponding value in direction 2 is slightly higher than 5000 (5100 read from Table 3.6). The approximate values for the *upper and lower quantiles* may also be observed from the quantile plots. Thus for example the lower quartile (0.25 quantile) in direction 1 is approximately equal to 4000 while the upper quartile (0.75 quantile) is close to 5000.

The slope of the quantile plot indicates the concentration of the data; a steep slope corresponds to a low concentration and a flat slope to a high concentration. The highest local concentration occurs when there are many observations with exactly the same value and this appears on the quantile plot by a horizontal series of points. For direction 1 the slope is quite flat up to about the 0.7 quantile. Thereafter the slope increases and thus the concentration of the data is smaller. This matches the information provided by the one-dimensional scatter plot in Fig. 3.2. The 0.7 quantile corresponds to a value close to 5000. It can be seen from Fig. 3.2 that for larger observed traffic flow values the concentration of the observations decreases.

Quantile plots may also provide information regarding the symmetry of data. If the observations in the data set are symmetrically dispersed around the median then the shape of the quantile plot in the upper half is a double mirrored image of the shape from the lower half. From Fig. 3.9 it can be seen that for both directions the data are not symmetric and that for direction 2 the asymmetry is more pronounced.

Following the same procedure as described above, the concrete cube compressive strength data are plotted in Fig. 3.10, against the respective quantile values given in Table 3.7. It can be seen that the quantile plot has an almost constant slope over the whole range of observations.

From Table 3.7 it can be seen that no observation corresponds directly to the median of the data set. In general the evaluation of a quantile which does not correspond to a given observation must be based on an interpolation. This may be performed by first calculating the hypothetical ith observation $\hat{x}_i^o$ equal to a given quantile Q_υ with the index υ:

$$\upsilon = \frac{i}{n+1} \tag{3.10}$$

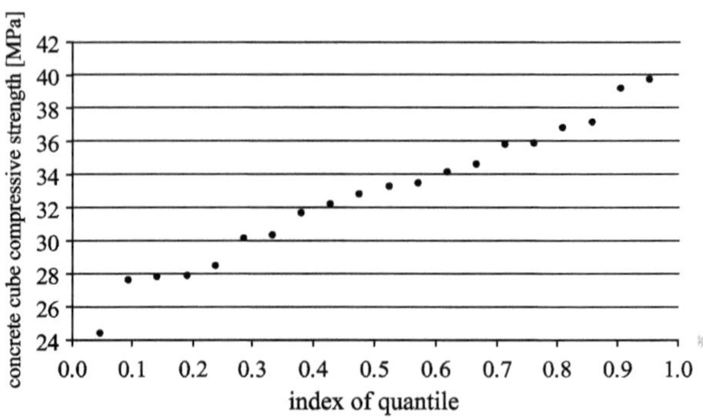

Fig. 3.10 Quantile plots of the observed concrete cube compressive strength

Solving Eq. 3.10 for i yields:

$$i = \upsilon(n+1) \tag{3.11}$$

If i is an integer, the ith observation $\hat{x}_i^o$ exists and is equal to Q_υ. If i is not an integer it will have a value consisting of an integer part, say k, and a fractional part, say p. The quantile $Q_\upsilon = \hat{x}_i^o$ using interpolation is given as:

$$\hat{x}_i^o = (1-p)\hat{x}_k^o + p\hat{x}_{k+1}^o \tag{3.12}$$

For example, in the case of the concrete cube compressive strength data (Table 3.7) looking for the upper quartile (0.75 quantile) gives a value of i equal to:

$$i = 20 \cdot 0.75 + 0.75 = 15.75$$

$$\text{thus } k = 15 \text{ and } p = 0.75.$$

Therefore, based on Eq. 3.12 the 0.75 quantile is:

$$0.25\hat{x}_{15}^o + 0.75\hat{x}_{16}^o = 0.25 \cdot 35.8 + 0.75 \cdot 35.9 = 35.875$$

3.3.4 Tukey Box Plots

Tukey box plots provide information about several sample characteristics of the observations contained in a data set, see Fig. 3.11.

The median is typically represented by a circle or a horizontal line within the box. The upper and lower sides of the box indicate the values of the upper and the lower quartiles, respectively. The distance between these quartiles is called the interquartile range, r; 50% of the data are located within this range. A large *interquartile range* indicates that the observations are widely dispersed around the median and vice versa.

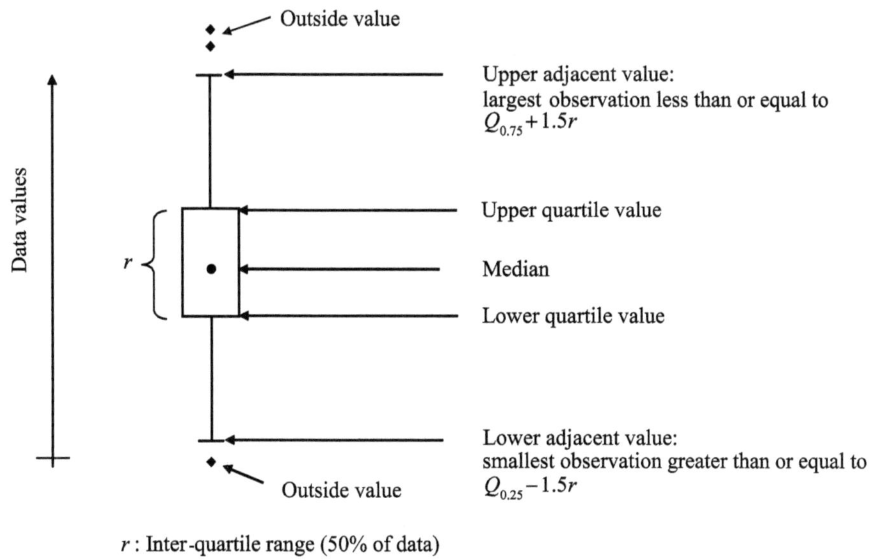

Fig. 3.11 Tukey box plot with indication of the characteristics of the data set

Table 3.8 Sample characteristics for the Tukey box plot for the traffic flow data in the Gotthard tunnel (Table 3.2)

Statistic	Direction 1	Direction 2
lower adjacent value	3087	3677
lower quartile	4085	4757
median	4419	5100
upper quartile	5001	5688
upper adjacent value	5892	6308
outside values	6551	7357
	7118	9323
	7974	10256
		11748

Another feature of the Tukey box plot is the *adjacent values*. The *upper adjacent value* is defined as the largest observation less than or equal to the upper quartile plus $1.5r$. The *lower adjacent value* is defined as the smallest observation greater than or equal to the lower quartile minus $1.5r$. If an observation has a value outside the adjacent values, the observation is called an *outside value* and is shown in the box plot by a single point.

In Table 3.8 the sample characteristics needed to construct the Tukey box plots for the traffic flow data are given. The Tukey box plot for this data is shown in Fig. 3.12.

The symmetry of the observations represented in a Tukey box plot may be partially assessed. From Fig. 3.12 can be seen that in direction 1 the observations with

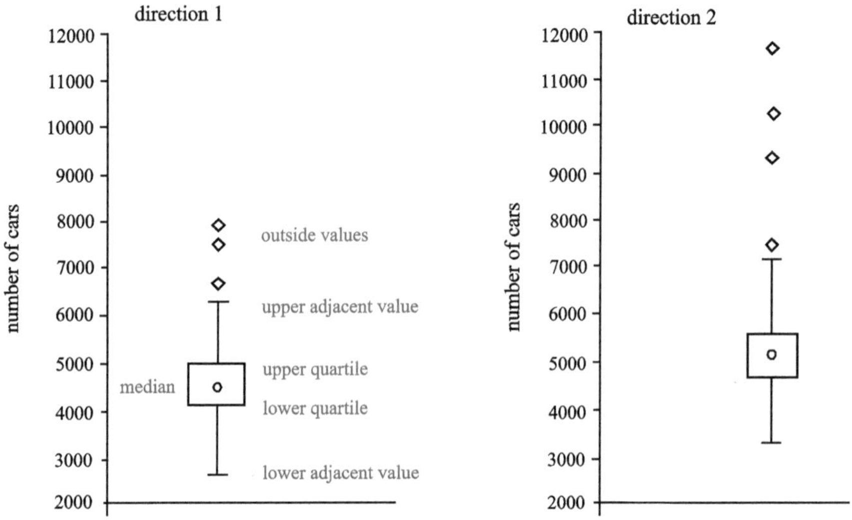

Fig. 3.12 Tukey box plots of the traffic flow data in the Gotthard tunnel

Fig. 3.13 Tukey box plot of the concrete cube compressive strength data [MPa]

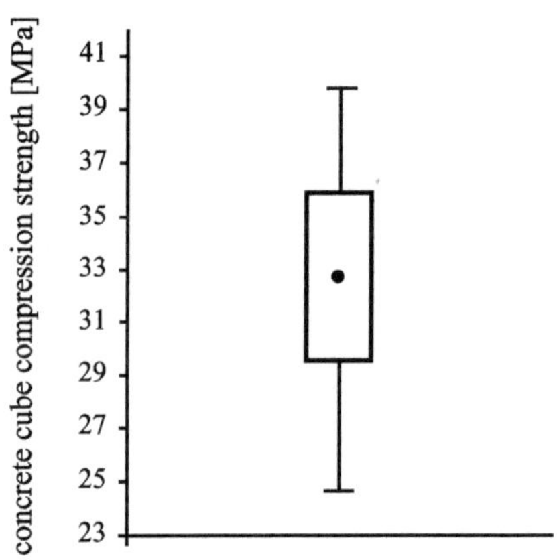

values in the lower and the upper range are more symmetric than the ones in direction 2. It is seen that the observed values of traffic flow in direction 2 are systematically larger than for direction 1.

In Fig. 3.13 the Tukey box plot for the concrete cube compressive strength data is given based on the evaluation of the respective sample statistics given in Table 3.9. For this set of data there are no outside values as the upper adjacent value is the maximum value of the data and the lower adjacent value corresponds to the lower value of the data.

Table 3.9 Statistics for the Tukey box plot for the concrete cube compressive strength data [MPa] (Table 3.1)

Statistic	Value
lower adjacent value	24.40
lower quartile	28.90
median	33.05
upper quartile	35.88
upper adjacent value	39.70

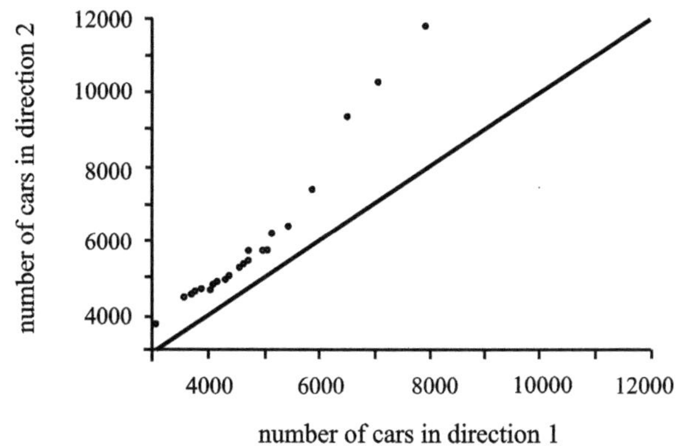

Fig. 3.14 Q-Q plot of the traffic flow observations in the two directions

3.3.5 Q-Q Plots and Tukey Mean-Difference Plot

Quantile-quantile plots (or in short Q-Q plots) provide an efficient means of comparing observations from different data sets. For example in the case of the traffic flow data a comparison may be made between the number of cars in one direction and the number of cars in another direction. To do so, the corresponding quantiles may be compared, i.e. the 0.25 quantile of the observations for direction 1 with the 0.25 quantile of the observations for direction 2 etc. For this purpose the corresponding quantiles are plotted against each other in a Q-Q plot.

The data sets as given in Table 3.6 contain the same number of observations and therefore their Q-Q plot may be made by plotting the evaluated quantiles against each other, see Fig. 3.14.

If the data sets compared do not have the same number of observations then the quantiles for the observations of one data set are evaluated first and subsequently the corresponding quantiles for the other data set are established by interpolation. From Fig. 3.14 it is seen that the traffic flow is higher over the full range of observations for direction 2 as compared to direction 1. If the Q-Q plot would result in a line close to the $y = x$ line, then the data would have nearly identical distributions.

Another graphical representation that facilitates the comparison of the observations contained in two different data sets is the *Tukey mean-difference plot*. Here $\hat{y}_i^o - \hat{x}_i^o$ is plotted against $(\hat{y}_i^o + \hat{x}_i^o)/2$, where $\hat{y}_i^o$ and $\hat{x}_i^o$ are the observations of the

Table 3.10 Values for the Tukey mean-difference plot of the traffic flow data in the Gotthard tunnel

Direction 1	Direction 2	$\hat{y}_i^o - \hat{x}_i^o$	$(\hat{y}_i^o + \hat{x}_i^o)/2$
3087	3677	590	3382.0
3578	4453	875	4015.5
3710	4480	770	4095.0
3737	4560	823	4148.5
3906	4635	729	4270.5
4029	4648	619	4338.5
4041	4672	631	4356.5
4085	4757	672	4421.0
4103	4791	688	4447.0
4164	4815	651	4489.5
4323	4880	557	4601.5
4359	4928	569	4643.5
4366	4946	580	4656.0
4368	5005	637	4686.5
4371	5013	642	4692.0
4419	5100	681	4759.5
4563	5220	657	4891.5
4588	5235	647	4911.5
4664	5281	617	4972.5
4667	5318	651	4992.5
4727	5398	671	5062.5
4739	5401	662	5070.0
4741	5679	938	5210.0
5001	5688	687	5344.5
5098	5729	631	5413.5
5193	6183	990	5688.0
5457	6308	851	5882.5
5892	7357	1465	6624.5
6551	9323	2772	7937.0
7118	10256	3138	8687.0
7974	11748	3774	9861.0

ordered data sets being compared. The evaluation of the means and differences for the traffic flow data is provided in Table 3.10.

In Fig. 3.15 the Tukey mean-difference plot is given for the traffic flow data.

The Tukey mean-difference plot indicates that there is a systematic difference of 600 cars per day for traffic situations corresponding to a mean traffic flow of up to 6000 cars per day. Thereafter, the differences in the two directions seem to be proportional in the mean traffic flow.

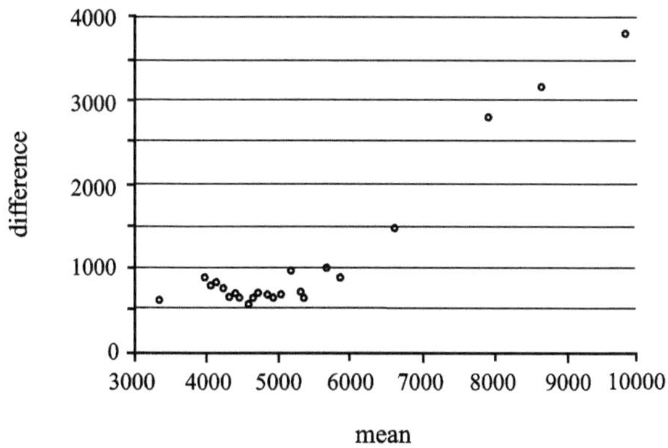

Fig. 3.15 Tukey mean-difference plot for the traffic flow data

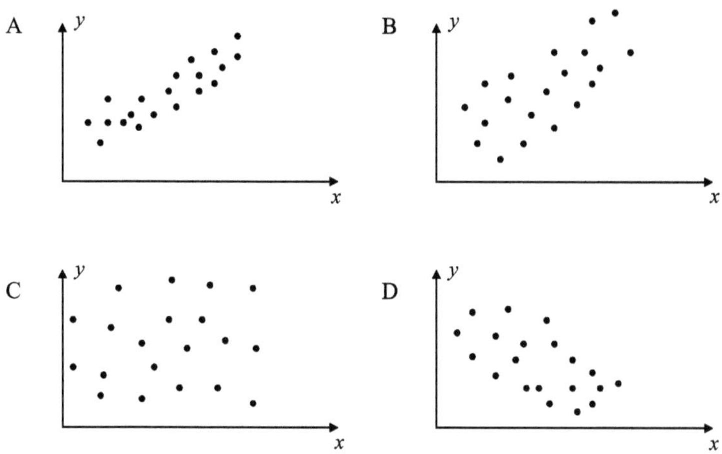

Fig. 3.16 Plotted data sets

3.4 Self Assessment Questions/Exercises

1. What is the purpose of descriptive statistics?
2. Within which interval can the coefficient of correlation of two data sets lie? What do the extreme values of the interval express?
3. What is the role of the interval width chosen for building up a histogram for the representation of a data set?
4. Which characteristics of a data set can be represented with a Tukey box plot?
5. How can Q-Q plots be used?
6. Provide a rough estimate of the correlation coefficient of the data sets plotted in Fig. 3.16.

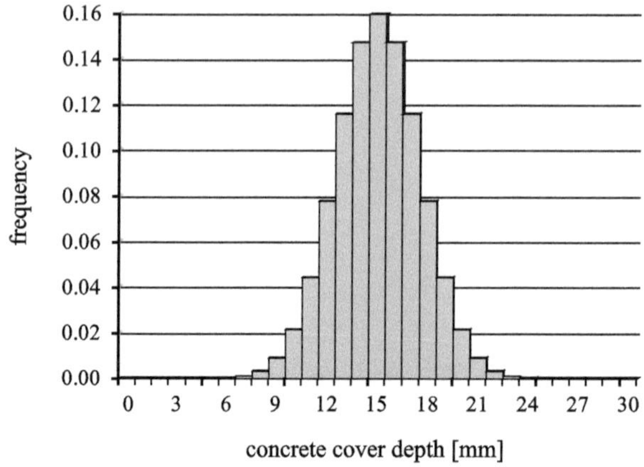

Fig. 3.17 Histogram of concrete cover depth measurements

Table 3.11 Statistical terms

	a	b	c
Mean	☐	☐	☐
Coefficient of variation	☐	☐	☐
Standard deviation	☐	☐	☐
Median	☐	☐	☐
Sample size	☐	☐	☐
Quartile	☐	☐	☐

A $r_{XY} \approx \cdots$ B $r_{XY} \approx \cdots$

C $r_{XY} \approx \cdots$ D $r_{XY} \approx \cdots$

7. A number of statistical terms are shown in Table 3.11. Check if the terms have something to do with (a) location parameter, (b) dispersion parameter or (c) none of the above.
8. Measurements were taken of the concrete cover depth of a bridge column. The histogram of the measured values has been plotted in Fig. 3.17.
   The sample mean is equal to 0.16 mm. ☐
   The sample mean is equal to 15 mm. ☐
   The mode of the data set is equal to 15 mm. ☐
9. Which of the following are features of a symmetrical frequency diagram?
   The variance is equal to the coefficient of variation. ☐
   The mode is equal to the median. ☐
   The skewness is equal to zero. ☐
   None of the above. ☐

Chapter 4
Uncertainty Modeling

Lecture 4 (Aim of the Present Lecture) The aim of the present lecture is to provide a fundamental understanding of uncertainty and how this affects engineering decision making. Furthermore, random variables are introduced and it is explained how they may be characterized depending on the given situation.

On the basis of the lecture it is expected that the reader will acquire knowledge and skills with regard to:

- Why do uncertainties influence engineering problems and decision making?
- Which are the principally different types of uncertainties?
- Why is it useful to differentiate between different types of uncertainties?
- Which types of uncertainties can be reduced?
- In what way can uncertainties depend on time?
- In what way can scale influence uncertainties?
- What is a random variable and how can it be characterized?
- How are cumulative distribution and probability density functions related?
- What is a discrete and what is a continuous probability distribution?
- How are the moments of a random variable defined?
- How is the expectation operation defined?

4.1 Introduction

A central role for engineers is to provide basis for decision making with regard to the cost efficient safeguarding of personnel, environment and assets in situations where uncertainties are at hand. A classical example is the decision problem of choosing the height of a dike. The risk of dike flooding can be reduced by increasing the height of the dike; however, due to the inherent natural variability in the water level

M.H. Faber, *Statistics and Probability Theory*,
Topics in Safety, Risk, Reliability and Quality 18,
DOI 10.1007/978-94-007-4056-3_4,

a certain probability of dike flooding in a given reference period will always remain. Risk assessment within the theoretical framework of decision analysis (introduced in Chap. 7) can help us in deciding on the optimal dike height by weighing the benefits of reduced dike flooding risks with the costs of increasing the dike height. However, a prerequisite for the risk assessment is that the means for assessing the probability of dike flooding are established, and this in turn requires that a probabilistic model for the future water level is available.

4.2 Uncertainties in Engineering Problems

For the purpose of discussing the phenomenon uncertainty in more detail, let us initially assume that the universe is deterministic and that our knowledge about the universe is perfect. This implies that it is possible by means of e.g. a set of exact equation systems and known boundary conditions by means of analysis to achieve perfect knowledge about any state, quantity or characteristic which otherwise cannot be directly observed or has yet not taken place. In principle, following this line of reasoning, the future as well as the past would be known or assessable with certainty. Considering the dike flooding problem it would thus be possible to assess the exact number of floods which would occur in a given reference period (the frequency of floods) for a given dike height and an optimal decision can be achieved by *cost benefit analysis*.

Whether the universe is deterministic or not is a rather deep philosophical question. Despite the obviously challenging aspects of this question its answer is, however, not a prerequisite for purposes of engineering decision making because even if the universe is deterministic our knowledge about it is still in part incomplete and/or uncertain.

In engineering decision analysis subject to uncertainties such as *Quantitative Risk Analysis* (QRA) and *Structural Reliability Analysis* (SRA) a commonly accepted view point is that *uncertainties* should be interpreted and differentiated with regard to their type and origin. In this way it has become standard to differentiate between uncertainties due to *inherent natural variability*, *model uncertainties* and *statistical uncertainties*. Whereas the first mentioned type of uncertainty is often denoted *aleatory* (or type 1) uncertainty, the two latter types are referred to as *epistemic* (or type 2) uncertainties. Without further discussion here, it is just stated that, in principle, all prevailing types of uncertainties should be taken into account in engineering decision analysis within the framework of *Bayesian probability theory*.

Considering again the dike example it can be imagined that an *engineering model* might be formulated where future extreme water levels are predicted in terms of a regression of previously observed annual extremes. In this case the uncertainty due to inherent natural variability would be the uncertainty associated with the annual extreme water level. The model chosen for the annual extreme water level events would by itself introduce model uncertainties and the parameters of the model would introduce statistical uncertainties as their estimation would be based on a limited

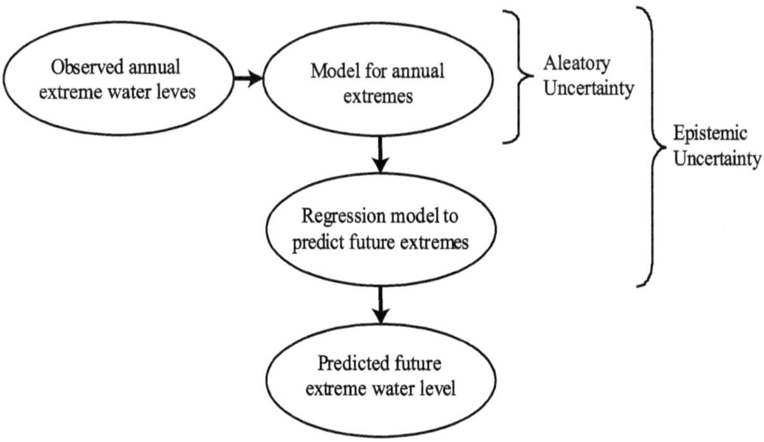

Fig. 4.1 Illustration of uncertainty composition in a typical engineering problem

number of observed annual extremes. Finally, the extrapolation of the annual extreme model to extremes over longer periods of time would introduce additional model uncertainties. The uncertainty associated with the future extreme water level is thus composed as illustrated in Fig. 4.1. Whereas the inherent natural variability is often understood as the uncertainty caused by the fact that the universe is not deterministic, it may also be interpreted simply as the uncertainty which cannot be reduced by means of collection of additional information. It is seen that this definition implies that the amount of uncertainty due to inherent natural variability depends on the models applied in the formulation of the engineering problem. Presuming that a refinement of models corresponds to looking in a more detailed manner at the problem at hand, one could say that the uncertainty structure influencing a problem is scale dependent.

Having formulated a model for the prediction of future extreme water levels and taking into account the various prevailing types of uncertainties, the probability of flooding within a given reference period can be assessed and, just as in the case of a deterministic and perfectly known universe, the optimum dike height can be decided, based on a cost-benefit assessment.

It is interesting to notice that the type of uncertainty associated with the state of knowledge has a time dependency. Following Fig. 4.2 it is possible to observe an uncertain phenomenon when it has occurred. In principle, if the observation is perfect without any errors the knowledge about the phenomenon is perfect. The modeling of the same phenomenon in the future, however, is uncertain as this involves models subject to natural variability, model uncertainty and statistical uncertainty. Often, but not always, the models available tend to lose their precision rather fast so that phenomena lying just a few days or weeks ahead can be predicted only with significant uncertainty. An extreme example of this concerns the prediction of the weather.

The above discussion shows another interesting effect, namely that the uncertainty associated with a model concerning the future, transforms from a mixture of

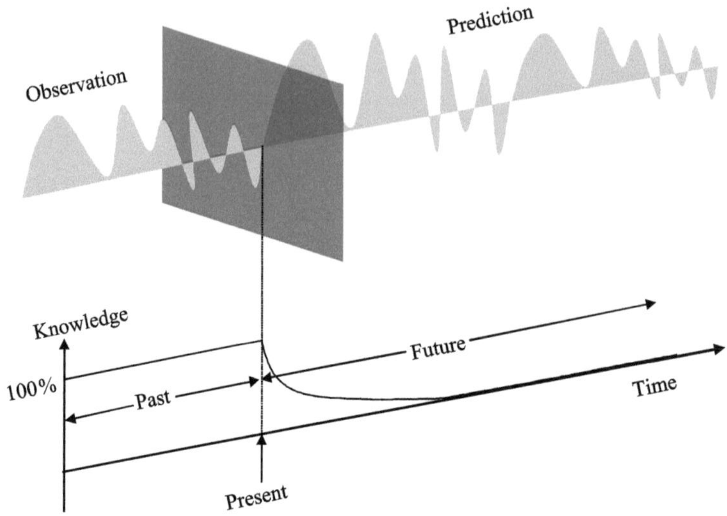

Fig. 4.2 Illustration of the time dependence of knowledge

aleatory and epistemic uncertainty to a purely epistemic uncertainty when the modeled phenomenon is observed. This transition of the type of uncertainty is significant because it facilitates that the uncertainty is reduced by utilization of updating based on observations.

4.3 Random Variables

The performance of an engineered system, facility or installation (in the following referred to as a system) may usually be modeled in mathematical and physical terms in conjunction with empirical relations.

For a given set of model parameters the performance of the considered system can be determined on the basis of this model. The basic *random variables* are defined as the parameters that represents the available knowledge as well as the associated uncertainty in the considered model.

The basic random variables must be able to represent all types of uncertainties that are included in the analysis. The uncertainties, which must be considered are, as previously mentioned, the physical uncertainty, the statistical uncertainty and the model uncertainty. The *physical uncertainties* are typically uncertainties associated with the loading environment, the geometry of the structure, the material properties and the repair qualities. The *statistical uncertainties* arise due to incomplete statistical information e.g. due to a small number of material tests. Finally, the model uncertainties must be considered to take into account the uncertainty associated with the idealized mathematical descriptions used to approximate the actual physical behavior of the structure.

Modern methods of reliability and risk analysis allow for a very general representation of these uncertainties ranging from non-stationary stochastic processes and fields to time invariant random variables, see e.g. Melchers [10]. In most cases it is sufficient to model the uncertain quantities by random variables with given cumulative distribution functions and distribution parameters estimated on basis of statistical and/or subjective information. Therefore the following is concerned with a basic description of the characteristics of random variables.

4.3.1 Cumulative Distribution and Probability Density Functions

A random variable, which can take on any value, is called a *continuous random variable*. The probability that such a random variable takes on a specific value is zero. The probability that a continuous random variable, X, is less than or equal to a value, x, is given by the cumulative distribution function:

$$F_X(x) = P(X \leq x) \tag{4.1}$$

In general, capital letters denote a random variable and small letters denote an outcome or realization of a random variable. An example of a continuous cumulative distribution function is illustrated in Fig. 4.3A).

For continuous random variables the *probability density function* is given by:

$$f_X(x) = \frac{dF_X(x)}{dx} \tag{4.2}$$

An example of a continuous probability density junction is illustrated in Fig. 4.3B).

The probability of an outcome in the interval $[x; x + dx]$ where dx is small, is given by $P(X \in [x; x + dx]) = f_X(x)\,dx$.

Random variables with a finite or infinite countable sample space are called discrete random variables. For *discrete random variables* the cumulative distribution function is given as:

$$P_X(x) = \sum_{x_i < x} p_X(x_i) \tag{4.3}$$

where $p_X(x_i)$ is the probability density function given as:

$$p_X(x_i) = P(X = x_i) \tag{4.4}$$

A discrete cumulative distribution function and probability density function is illustrated in Fig. 4.4.

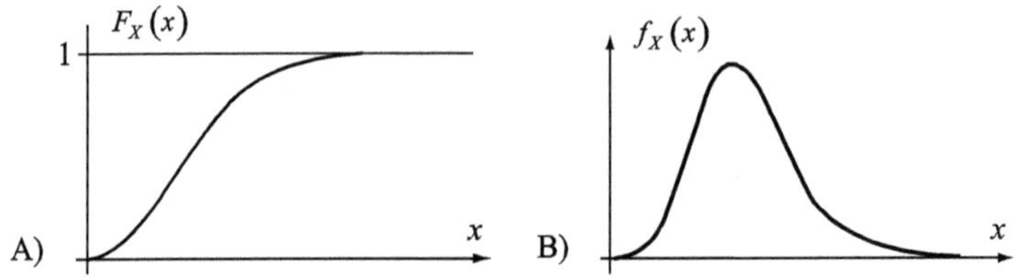

Fig. 4.3 Illustration of **A)** a cumulative distribution function and **B)** a probability density function for a continuous random variable

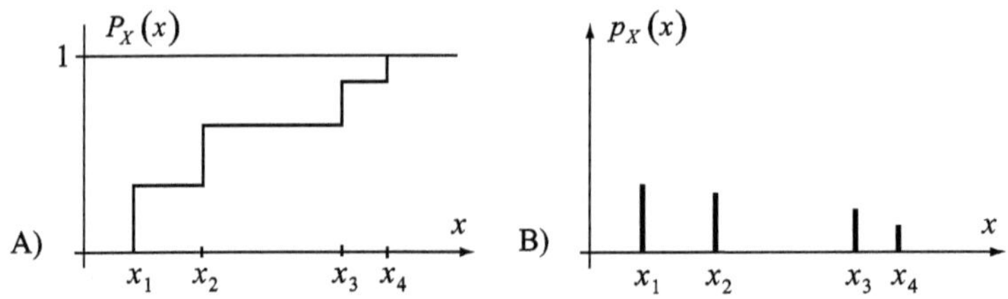

Fig. 4.4 Illustration of **A)** a cumulative distribution function and **B)** a probability density function for a discrete random variable

4.3.2 Moments of Random Variables and the Expectation Operator

Probability distributions may be defined in terms of their *parameters* or *moments*. Often cumulative distribution functions and probability density functions are written as $F_X(x; \mathbf{p})$ and $f_X(x; \mathbf{p})$ respectively to indicate the parameters $\mathbf{p}$ (or moments) defining the functions. Whether the cumulative distribution and density function are defined by their moments or by their parameters is a matter of convenience and it is generally possible to establish one from the other.

The ith moment λ_i of a continuous random variable is defined by:

$$\lambda_i = \int_{-\infty}^{\infty} x^i f_X(x)\, dx \tag{4.5}$$

and for a discrete random variable by:

$$\lambda_i = \sum_{j=1}^{n} x_j^i p_X(x_j) \tag{4.6}$$

The *mean* (or *expected value*) of a continuous random variables X is defined as the first moment:

$$\mu_X = E[X] = \int_{-\infty}^{\infty} x\, f_X(x)\, dx \tag{4.7}$$

For a discrete random variable S, the mean is given as:

$$\mu_X = E[X] = \sum_{j=1}^{n} x_j p_X(x_j) \tag{4.8}$$

where $E[\cdot]$ denotes the expectation operator.

Similarly the *variance*, ${\sigma_X}^2$, is described by the *second central moment*, i.e. for continuous random variables it is:

$${\sigma_X}^2 = Var[X] = E\left[(X - \mu_X)^2\right] = \int_{-\infty}^{\infty} (x - \mu_X)^2 f_X(x)\,dx \tag{4.9}$$

and for discrete random variables it is:

$${\sigma_X}^2 = Var[X] = \sum_{j=1}^{n} (x_j - \mu_X)^2 p_X(x_j) \tag{4.10}$$

where $Var[X]$ denotes the variance of X.

The ratio between the standard deviation σ_X and the expected value μ_X of a random variable X is denoted the *coefficient of variation* V_X and is given by:

$$V_X = \frac{\sigma_X}{\mu_X} \tag{4.11}$$

The coefficient of variation provides a useful descriptor for the variability of a random variable around its expected value.

4.3.3 Example 4.1—Uniform Distribution

As an example consider a continuous random variable with a uniform (constant) probability density function in the interval $[a; b]$ as illustrated in Fig. 4.5.

The probability density function for a *uniformly distributed random variable X* is easily seen to be:

$$f_X(x; a, b) = \begin{cases} 0, & x < a \\ \frac{1}{b-a}, & a \leq x \leq b \\ 0, & x > b \end{cases} \tag{4.12}$$

remembering that the area under the probability density function must integrate to 1. In Eq. 4.12, a and b are the parameters of the probability density function.

The cumulative distribution function for a uniformly distributed random variable X is thus:

$$F_X(x; a, b) = \begin{cases} 0, & x < a \\ \int_a^x f_X(y; a, b)\,dy = \int_a^x \frac{1}{b-a}\,dy = \frac{(x-a)}{(b-a)}, & a \leq x \leq b \\ 1, & x > b \end{cases} \tag{4.13}$$

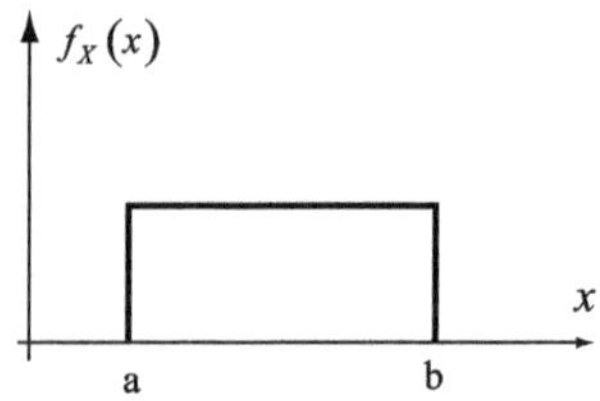

Fig. 4.5 Continuous random variable with a uniform density function

The first moment i.e. the mean value (see Eq. 4.7) of a continuous random variable X with uniform distribution is thus:

$$\mu_X = E[X] = \int_a^b x f_X(x; a, b)\, dx = \int_a^b \frac{x}{b-a}\, dx = \left. \frac{x^2}{2(b-a)} \right|_a^b$$
$$= \frac{(b+a)}{2} \tag{4.14}$$

and the variance σ_X^2 (see Eq. 4.9) is given through the second central moment:

$$\sigma_X^2 = E\left[(X-\mu_X)^2\right] = \int_a^b (x-\mu_X)^2 f_X(x; a, b)\, dx = \int_a^b \frac{(x-\mu_X)^2}{(b-a)}\, dx$$
$$= \left. \frac{\frac{1}{3}x^3 - x^2\mu_X + x{\mu_X}^2}{(b-a)} \right|_a^b$$
$$= \frac{1}{12}(b-a)^2 \tag{4.15}$$

Lecture 5 (Aim of the Present Lecture) The aim of the present lecture is to introduce the properties associated with the main characteristics of vectors of random variables and how to assess these characteristics. Furthermore, it is described how probabilistic characterizations of functions of random variables can be established.

On the basis of the lecture it is expected that the reader will acquire knowledge and skills with regard to:

- How can the expectation operation be performed on a linear combination of random variables?
- How can the expectation operation be performed on a linear combination of functions of random variables?
- Which rule applies for the expectation operation of functions of random variables?

- What is the relation between the expectation operation and the variance operation?
- Which are the properties of the expectation and the variance operator?
- What is a random vector and what is a joint moment?
- How is the covariance between two random variables defined?
- How is the correlation coefficient defined and what information does it contain?
- What is a marginal probability distribution?
- What is a conditional probability distribution?
- How can the probability distribution for the sum of two random variables be established?
- How can the probability distribution for a function of random variables be established?

4.3.4 Properties of the Expectation Operator

It is useful to note that the *expectation operation* possesses the following properties, where a, b and c are constants and X is a random variable:

$$\begin{aligned} E[c] &= c \\ E[cX] &= cE[X] \\ E[a+bX] &= a+bE[X] \\ E[g_1(X)+g_2(X)] &= E[g_1(X)]+E[g_2(X)] \end{aligned} \tag{4.16}$$

The implication of the last equation is that expectation, like differentiation or integration, is a linear operation. This linearity property is useful since it can be used, for example, to find the following formula for the variance of a random variable X in terms of more easily calculated quantities:

$$\begin{aligned} Var[X] &= E\left[(X-\mu_X)^2\right] = E\left[X^2+\mu_X^2-2\mu_X X\right] = \mu_X^2 + E\left[X^2\right] - 2\mu_X E[X] \\ &= \mu_X^2 + E\left[X^2\right] - 2\mu_X^2 = E\left[X^2\right] - \mu_X^2 \end{aligned} \tag{4.17}$$

By application of Eq. 4.17 the following properties of the *variance operator* $Var[\cdot]$ can easily be derived:

$$\begin{aligned} Var[c] &= 0 \\ Var[cX] &= c^2 Var[X] \\ Var[a+bX] &= b^2 Var[X] \end{aligned} \tag{4.18}$$

where a, b and c are constants and X is a random variable.

From Eq. 4.17 it is furthermore seen that in general it is $E[g(X)] \neq g(E[X])$. In fact for convex functions $g(x)$ it can be shown that the following inequality is valid (*Jensen's inequality*):

$$E[g(X)] \geq g(E[X]) \tag{4.19}$$

where the equality holds if $g(X)$ is linear.

4.3.5 Random Vectors and Joint Moments

If a n-dimensional vector of continuous random variables $\mathbf{X} = (X_1, X_2, \ldots, X_n)^T$, is considered the *joint cumulative distribution function* is given by:

$$F_{\mathbf{X}}(\mathbf{x}) = P\left(X_1 \leq x_1 \cap X_2 \leq x_2 \cap \cdots \cap X_n \leq x_n\right) \tag{4.20}$$

and the *joint probability density function* is:

$$f_{\mathbf{X}}(\mathbf{x}) = \frac{\partial^n}{\partial x_1 \partial x_2 \cdots \partial x_n} F_{\mathbf{X}}(\mathbf{x}) \tag{4.21}$$

An example of a joint probability density function is illustrated in Fig. 4.6 in the form of a binormal distribution.

The *covariance* $C_{X_iX_j}$ between X_i and X_j is defined by:

$$\begin{aligned} C_{X_iX_j} &= Cov[X_i, X_j] = E\left[(X_i - \mu_{X_i})(X_j - \mu_{X_j})\right] \\ &= \int_{-\infty}^{\infty}\int_{-\infty}^{\infty} (x_i - \mu_{X_i})(x_j - \mu_{X_j}) f_{X_iX_j}(x_i, x_j)\, dx_i\, dx_j \end{aligned} \tag{4.22}$$

and is also called the *joint central moment* between the variables X_i and X_j.

The covariance expresses the linear dependence between two variables. It is evident that $C_{X_iX_i} = Var[X_i]$. On the basis of the covariance the *correlation coefficient* is defined by:

$$\rho_{X_iX_j} = \frac{C_{X_iX_j}}{\sigma_{X_i}\sigma_{X_j}} \tag{4.23}$$

It is seen that $\rho_{X_iX_i} = 1$. The correlation coefficients can only take values in the interval $[-1; 1]$.

A negative correlation coefficient between two random variables implies that if the outcome of one variable is large compared to its mean value, the outcome of the other variable is likely to be small compared to its mean value. A positive correlation coefficient between two variables implies that if the outcome of one variable is large compared to its mean value, the outcome of the other variable is also likely to be large compared to its mean value. If two variables are independent their correlation coefficient is zero and the *joint density function* is the product of the one-dimensional density functions. In many cases it is possible to obtain a sufficiently

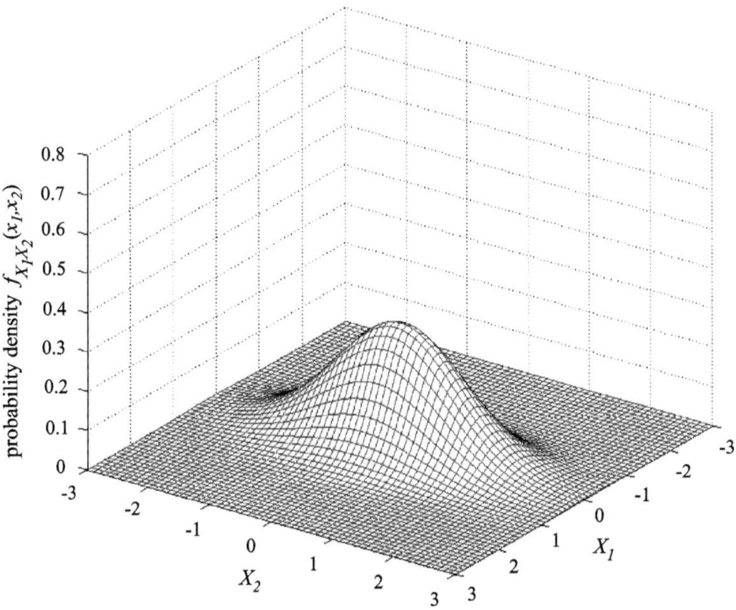

Fig. 4.6 Joint probability density function (binormal distribution)

accurate approximation to the *n-dimensional cumulative distribution function* from the one-dimensional distribution functions of the n variables and their parameters, and the correlation coefficients.

Finally using Eqs. 4.17, 4.18 and 4.22 it can be shown that the expected value $E[Y]$ and the variance $Var[Y]$, where Y is a linear function of the random vector $\mathbf{X} = (X_1, X_2, \ldots, X_n)^T$ i.e.:

$$Y = a_0 + \sum_{i=1}^{n} a_i X_i \tag{4.24}$$

are given by:

$$\begin{aligned} E[Y] &= a_0 + \sum_{i=1}^{n} a_i E[X_i] \\ Var[Y] &= \sum_{i=1}^{n} a_i^2 Var[X_i] + 2\left(\sum_{i=1}^{n} \sum_{j=i+1}^{n} a_i a_j C_{X_i X_j}\right) \end{aligned} \tag{4.25}$$

4.3.6 Example 4.2—Linear Combinations and Random Variables

Consider the linear function $Y = 5 + 2X_1 + 3X_2$. The expected values of the random variables X_1 and X_2 are $E[X_1] = 7$ as well as $E[X_2] = 4$. The variances are

$Var[X_1] = 1$ and $Var[X_2] = 1.5$. Both random variables X_1 and X_2 are assumed to be independent, hence the covariance is $C_{X_1X_2} = 0$.

The expected value $E[Y]$ is determined as follows:

$$E[Y] = a_0 + \sum_{i=1}^{n} a_i E[X_i] = a_0 + a_1 E[X_1] + a_2 E[X_2]$$
$$= 5 + 2E[X_1] + 3E[X_2] = 5 + 2 \cdot 7 + 3 \cdot 4 = 31$$

The variance $Var(Y)$ is determined as follows:

$$Var[Y] = \sum_{i=1}^{n} a_i^2 Var[X_i] + 2\left(\sum_{i=1}^{n} \sum_{j=i+1}^{n} a_i a_j C_{X_i X_j}\right) = a_1^2 Var[X_1] + a_2^2 Var[X_2]$$
$$= 2^2 Var[X_1] + 3^2 Var[X_2] = 4 \cdot 1 + 9 \cdot 1.5 = 17.5$$

4.3.7 Conditional Distributions and Conditional Moments

The *conditional probability density function* for the random variable X_1, conditional on the outcome of the random variable X_2 is denoted $f_{X_1|X_2}(x_1|x_2)$ and defined by:

$$f_{X_1|X_2}(x_1|x_2) = \frac{f_{X_1,X_2}(x_1,x_2)}{f_{X_2}(x_2)} \tag{4.26}$$

in accordance with the definition of conditional probability given previously. Figure 4.7 illustrates the conditional probability density function $f_{X_1|X_2}(x_1|x_2 = 0.5t)$ of the binormal distribution illustrated in Fig. 4.6.

As for the case when probabilities of events were considered two random variables X_1 and X_2 are said to be independent when there is:

$$f_{X_1|X_2}(x_1|x_2) = f_{X_1}(x_1) \tag{4.27}$$

By integration of Eq. 4.26 the conditional cumulative distribution $F_{X_1|X_2}(x_1|x_2)$ is obtained as:

$$F_{X_1|X_2}(x_1|x_2) = \frac{\int_{-\infty}^{x_1} f_{X_1,X_2}(z,x_2)\,dz}{f_{X_2}(x_2)} \tag{4.28}$$

and finally by integration of Eq. 4.28 weighed with the probability density function of X_2, i.e. $f_{X_2}(x_2)$ the unconditional cumulative distribution $F_{X_1}(x_1)$ is achieved by the *total probability theorem*:

$$F_{X_1}(x_1) = \int_{-\infty}^{\infty} F_{X_1|X_2}(x_1|x_2) f_{X_2}(x_2)\,dx_2 \tag{4.29}$$

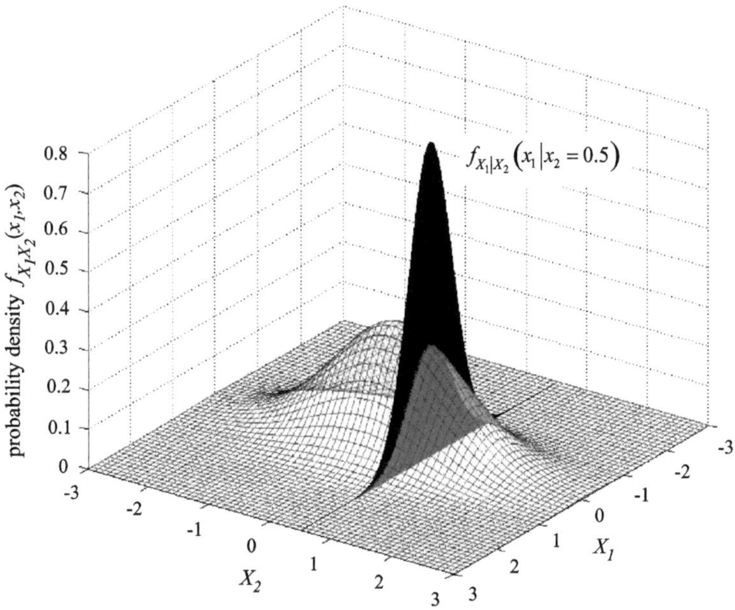

Fig. 4.7 Conditional probability density function

The *conditional moments* of jointly distributed continuous random variables follow straightforwardly from Eq. 4.7 by use of Eq. 4.27 and the conditional expected value $\mu_{X_1|X_2}$ and the conditional variance of e.g. the jointly distributed random variables X_1 given X_2 are evaluated by:

$$\begin{aligned} \mu_{X_1|X_2} &= E[X_1|X_2=x_2] = \int_{-\infty}^{\infty} x_1 f_{X_1|X_2}(x_1|x_2)\,dx_1 \\ (\sigma_{X_1|X_2})^2 &= E\big[(X_1-\mu_{X_1|X_2})^2|X_2=x_2\big] \\ &= \int_{-\infty}^{\infty} (x_1-\mu_{X_1|X_2})^2 f_{X_1|X_2}(x_1|x_2)\,dx_1 \end{aligned} \quad (4.30)$$

4.3.8 The Probability Distribution for the Sum of Two Random Variables

Based on the result in Eq. 4.26 the probability density function for the random variable $Y = X_1 + X_2$ may be derived for a given joint probability density function $f_{X_1,X_2}(x_1,x_2)$. First the conditional probability density function of Y given $X_1 = x_1$ is considered i.e.:

$$Y = x_1 + X_2 \quad \Leftrightarrow \quad X_2 = Y - x_1 \quad (4.31)$$

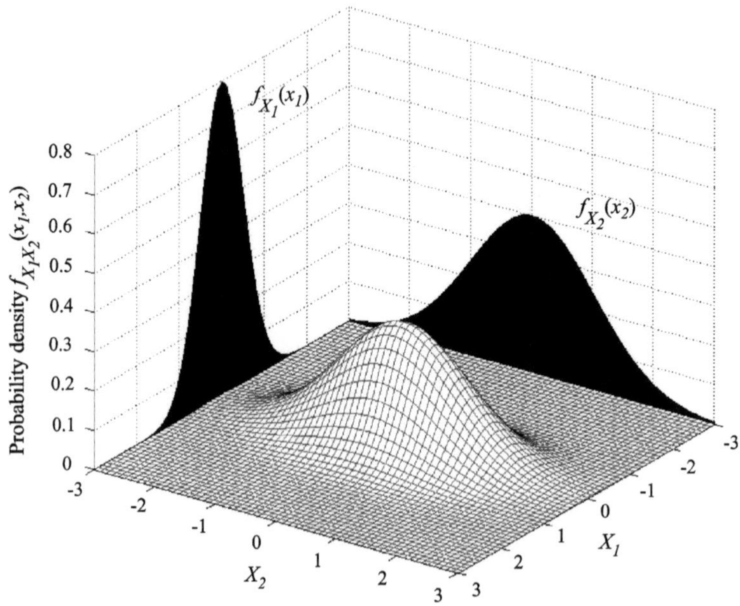

Fig. 4.8 The marginal probability density functions for a binormal distribution

where the conditional probability density function of X_2 given $X_1 = x_1$ is:

$$f_{X_2|X_1}(x_2|x_1) = \frac{f_{X_1,X_2}(x_1,x_2)}{f_{X_1}(x_1)} \tag{4.32}$$

thus the probability density function for Y given $X_1 = x_1$ can be written as:

$$f_{Y|X_1}(y|x_1) = f_{X_2|X_1}(y - x_1|x_1) \tag{4.33}$$

and the joint probability density function for $f_{Y,X_1}(y,x_1)$:

$$f_{Y,X_1}(y,x_1) = f_{X_2|X_1}(y - x_1|x_1) f_{X_1}(x_1) = f_{X_2,X_1}(y - x_1, x_1) \tag{4.34}$$

from which one can get the *marginal probability density function* of Y by integrating out over the definition domain of x_1 i.e.:

$$f_Y(y) = \int_{-\infty}^{\infty} f_{X_2,X_1}(y - x_1, x_1)\, dx_1 \tag{4.35}$$

Figure 4.8 illustrates both the marginal probability density functions of a binormal distribution.

For the special case where the variables X_1 and X_2 are independent, Eq. 4.35 can be written in the form of a *convolution integral*:

$$f_Y(y) = \int_{-\infty}^{\infty} f_{X_2}(y - x_1) f_{X_1}(x_1)\, dx_1 \tag{4.36}$$

4.3.9 Example 4.3—Density Function for the Sum of Two Random Variables—Special Case Normal Distribution

The total load L on a structural element, consisting of a live load N and self-weight G and expressed as $L = N + G$ is to be determined. Both the random variables—live load N and self weight G are assumed to follow the Normal distribution and to be independent—hence the covariance is $C_{GN} = 0$.

With the aid of the convolution integral it is possible to show that a sum of two Normal distributions follows a Normal distribution (see also the section on central limit theorem on page 62):

$$
\begin{aligned}
f_L(l) &= \int_{-\infty}^{\infty} f_G(l-n) f_N(n)\, dn \\
&= \int_{-\infty}^{\infty} \frac{1}{\sigma_G\sqrt{2\pi}} \exp\left(-\frac{1}{2}\left(\frac{l-n-\mu_G}{\sigma_G}\right)^2\right) \frac{1}{\sigma_N\sqrt{2\pi}} \exp\left(-\frac{1}{2}\left(\frac{n-\mu_N}{\sigma_N}\right)^2\right) dn \\
&= \frac{1}{\sqrt{\sigma_G^2+\sigma_N^2}\sqrt{2\pi}} \exp\left(-\frac{1}{2}\frac{(l-(\mu_N+\mu_G))^2}{\sqrt{\sigma_G^2+\sigma_N^2}}\right) \\
&= \frac{1}{\sigma_L\sqrt{2\pi}} \exp\left(-\frac{1}{2}\frac{(l-(\mu_L))^2}{\sigma_L}\right)
\end{aligned}
\tag{4.37}
$$

The convolution of two Normal distributions can be assessed by Eq. 4.25. The expected values and variances of workload N and self-weight G are:

$$
E[N] = 1\ \text{kN/m}^2, \qquad Var[G] = 0.2\ \left(\text{kN/m}^2\right)^2,
$$
$$
E[G] = 3\ \text{kN/m}^2, \qquad Var[G] = 0.3\ \left(\text{kN/m}^2\right)^2
$$

The expected value of the total load L is determined by:

$$
E[L] = 0 + 1 \cdot E[G] + 1 \cdot E[N] = 3 + 1 = 4\ \text{kN/m}^2
$$

The variance of the total load L is determined by:

$$
Var[N] = 1^2 \cdot Var[G] + 1^2 \cdot Var[N] + 0 = 0.3^2 + 0.2^2 = 0.13\ \left(\text{kN/m}^2\right)^2
$$

The density functions of the three random variables N, G and L are illustrated in Fig. 4.9.

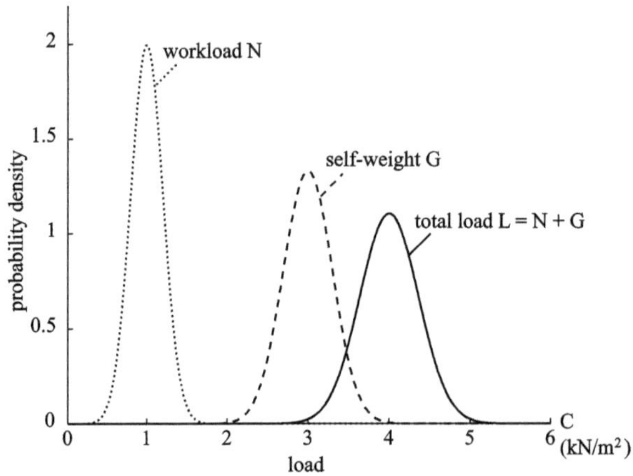

Fig. 4.9 Sum of two random variables

4.3.10 The Probability Distribution for Functions of Random Variables

In some cases it is interesting to be able to derive the cumulative distribution function $F_Y(y)$ for a random variable Y which is given as a function of another random variable X i.e. $Y = g(X)$, with given cumulative distribution function $F_X(x)$. Under the condition that the function $g(x)$ is monotonically increasing and furthermore, represents a one-to-one mapping of x into y, a realization of Y is only smaller than y_0 if correspondingly the realization of X is smaller than x_0 which in turn is given by $x_0 = g^{-1}(y_0)$. In this case the cumulative distribution function $F_Y(y)$ can be readily determined by:

$$F_Y(y) = P(Y \leq y) = P\big(X \leq g^{-1}(y)\big) \tag{4.38}$$

which is also written as:

$$F_Y(y) = F_X\big(g^{-1}(y)\big) \tag{4.39}$$

In accordance with Eq. 4.2 the probability density function $f_Y(y)$ is simply given by:

$$f_Y(y) = \frac{dF_X(g^{-1}(y))}{dy} \tag{4.40}$$

which immediately leads to:

$$f_Y(y) = \frac{dg^{-1}(y)}{dy} f_X\big(g^{-1}(y)\big) \tag{4.41}$$

and:

$$f_Y(y) = \frac{dx}{dy} f_X(x) \tag{4.42}$$

It is noticed that the application of Eqs. 4.41 and 4.42 necessitates that $g(x)$ is at least one time differentiable in regard to x.

Now if the function $g(x)$ instead of being monotonically increasing is monotonically decreasing, a realization of Y smaller that y_0 corresponds to a realization of X larger than x_0, in which case it is necessary to change the sign of the derivative dx/dy in Eq. 4.42. Generally, for monotonically increasing or decreasing one-to-one functions $g(x)$ there is:

$$f_Y(y) = \left|\frac{dx}{dy}\right| f_X(x) \tag{4.43}$$

As shown in e.g. Thoft-Christensen and Baker [13] the relationship given in Eq. 4.34 can be generalized to consider the case of jointly distributed random variables.

Consider the random vector $\mathbf{Y} = (Y_1, Y_2, \ldots, Y_n)^T$ with individual components given as one-to-one mapping monotonically increasing or decreasing functions g_i, $i = 1, 2, \ldots, n$ of the components of the random vector $\mathbf{X} = (X_1, X_2, \ldots, X_n)^T$ as:

$$Y_i = g_i(\mathbf{X}) \tag{4.44}$$

then there is:

$$f_{\mathbf{Y}}(\mathbf{y}) = |\mathbf{J}| f_{\mathbf{X}}(\mathbf{x}) \tag{4.45}$$

where $|\mathbf{J}|$ is the numerical value of the determinant of $\mathbf{J}$ given by:

$$\mathbf{J} = \begin{bmatrix} \frac{\partial x_1}{\partial y_1} & \cdots & \frac{\partial x_1}{\partial y_n} \\ \vdots & \ddots & \vdots \\ \frac{\partial x_n}{\partial y_1} & \cdots & \frac{\partial x_n}{\partial y_n} \end{bmatrix} \tag{4.46}$$

Finally, the expected value $E[Y]$ of a function $g(\mathbf{X})$ of the random vector $\mathbf{X} = (X_1, X_2, \ldots, X_n)^T$ is given by:

$$E[Y] = \int_{-\infty}^{\infty} \cdots \int_{-\infty}^{\infty} g(\mathbf{x}) f_{\mathbf{X}}(\mathbf{x})\, dx_1 \cdots dx_n \tag{4.47}$$

4.3.11 Example 4.4—Probability Distribution for a Function of Random Variables

The costs C of a project consist of fixed costs of 10,000 CHF and variable costs of 100 CHF for each working hour H, that is $C = 10,000 + 100H$. For the random

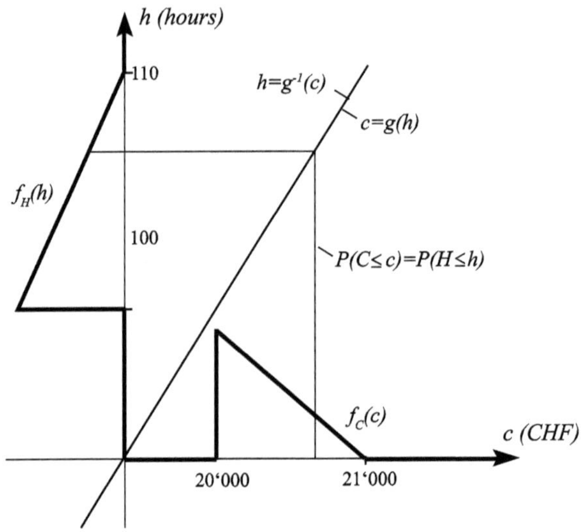

Fig. 4.10 Schematic representation of the distribution of a random variable H and the derived distribution for a random variable C

variable H, the probability distribution is given. How can the probability distribution for C be derived?

The relationship of the two random variables is monotonically increasing and one-to-one. The costs C are less than any particular value c only if the number of hours H is less than a particular value h, that is $F_C(c) = P[C \leq c] = P[H \leq g^{-1}(c)] = F_H(g^{-1}(c))$ (Fig. 4.10).

With $c = g(h) = 10,000 + 100h$ and $h = g^{-1}(c) = (c - 10,000)/100$ for this example we get:

$$F_C(c) = F_H\left(\frac{c - 10,000}{100}\right)$$

By applying Eq. 4.43, the probability density function can be established. By inserting $\frac{dh}{dc} = \frac{dg^{-1}(c)}{dc} = \frac{d(c-10,000)/100}{dc} = \frac{1}{100}$ we get:

$$f_C(c) = \frac{dh}{dc} f_H(h) = \frac{dg^{-1}(c)}{dc} f_H\left(g^{-1}(c)\right) = \frac{1}{100} f_H\left(\frac{c - 10,000}{100}\right)$$

For a probability density function $f_H(h) = 2.2 - 0.02h$ in the range of $100 \leq h \leq 110$, the following probability density function for $f_C(c)$ is obtained:

$$f_C(c) = \frac{1}{100}(2.2 - 0.02)\left(\frac{c - 10,000}{100}\right) = 0.042 - 0.000,002c$$

$$20,000 \leq c \leq 21,000$$

The range values for c is obtained from the function $c = 10,000 + 100h$.

In this simple case with a linear function, the shape of the density function does not change, as can be seen in Fig. 4.10. As an example of a case where the shape of the probability density function changes, consider the function $Q = ke^{\lambda T}$. Q, the number of bacteria in a tank, can increase starting from an initial population size k with a growth rate λ, until the tank is emptied after a random time period T.

Lecture 6 (Aim of the Present Lecture) The aim of the present lecture is to first summarize typical probability distribution functions applied in engineering uncertainty modeling. Thereafter, it is outlined how the Normal and the Lognormal probability distributions may be derived on the basis of the central limit theorem. Furthermore, as an introduction on how to model uncertain phenomena with random variability over time, random sequences and their characterization are introduced.
On the basis of the lecture it is expected that the reader will acquire knowledge and skills with regard to:

- What does the central limit theorem say?
- What is a standardized random variable?
- How can the variance of a linear combination of correlated Normal distributed random variables be calculated?
- How can the Lognormal distribution be derived?
- In what way can uncertain phenomena depend on "time"?
- What is a random sequence?
- What is a Bernoulli trial and what does it describe?
- For what can the Binomial distribution be used?
- What is a Geometric distribution and where can it be applied?

4.3.12 Probability Density and Distribution Functions

In Table 4.1 a selection of probability density and cumulative distribution functions is given with the definition of their distribution parameters and moments. With a given expectation operator and standard deviation, the parameters of the distributions can be assessed and the corresponding distributions can be drawn. In case of a non-shifted lognormal or exponential distribution, ε is equal to zero.

Illustrations for the distribution types given in Table 4.1 are shown in Fig. 4.11.

The relevance of the different distribution functions given in Table 4.1 in connection with the probabilistic modeling of uncertainties in engineering risk and reliability analysis is strongly case dependent and the reader is suggested to consult the application specific literature for specific guidance. In the following, however,

Table 4.1 Probability distributions, Schneider [11]

Distribution type	Parameter	Mean and standard deviation
Uniform distribution, $a \le x \le b$		
$f_X(x) = \frac{1}{b-a}$	a	$\mu = \frac{a+b}{2}$
$F_X(x) = \frac{x-a}{b-a}$	b	$\sigma = \frac{b-a}{\sqrt{12}}$
Normal distribution		
$f_X(x) = \frac{1}{\sigma\sqrt{2\pi}} \exp\left(-\frac{1}{2}\left(\frac{x-\mu}{\sigma}\right)^2\right)$	μ	μ
$F_X(x) = \frac{1}{\sigma\sqrt{2\pi}} \int_{-\infty}^{x} \exp\left(-\frac{1}{2}\left(\frac{t-\mu}{\sigma}\right)^2\right) dt$	$\sigma > 0$	σ
Shifted Lognormal, $x > \varepsilon$		
$f_X(x) = \frac{1}{(x-\varepsilon)\zeta\sqrt{2\pi}} \exp\left(-\frac{1}{2}\left(\frac{\ln(x-\varepsilon)-\lambda}{\zeta}\right)^2\right)$	λ	$\mu = \varepsilon + \exp\left(\lambda + \frac{\zeta^2}{2}\right)$
$F_X(x) = \Phi\left(\frac{\ln(x-\varepsilon)-\lambda}{\zeta}\right)$	$\zeta > 0$	$\sigma = \exp\left(\lambda + \frac{\zeta^2}{2}\right)\sqrt{\exp(\zeta^2) - 1}$
	ε	
Shifted Exponential, $x \ge \varepsilon$		
$f_X(x) = \lambda \exp(-\lambda(x - \varepsilon))$	ε	$\mu = \varepsilon + \frac{1}{\lambda}$
$F_X(x) = 1 - \exp(-\lambda(x - \varepsilon))$	$\lambda > 0$	$\sigma = \frac{1}{\lambda}$
Gamma, $x \ge 0$		
$f_X(x) = \frac{b^p}{\Gamma(p)} \exp(-bx)x^{p-1}$	$p > 0$	$\mu = \frac{p}{b}$
$F_X(x) = \frac{\Gamma(bx,p)}{\Gamma(p)}$	$b > 0$	$\sigma = \frac{\sqrt{p}}{b}$
where:		
$\Gamma(bx, p) = \int_0^{bx} t^{p-1} \exp(-t)\, dt$		
$\Gamma(p) = \int_0^{\infty} t^{p-1} \exp(-t)\, dt$		
Beta, $a \le x \le b$		
$f_X(x) = \frac{\Gamma(r+t)}{\Gamma(r)\Gamma(t)} \frac{(x-a)^{r-1}(b-x)^{t-1}}{(b-a)^{r+t-1}}$	a	$\mu = a + (b-a)\frac{r}{r+t}$
$F_X(x) = \frac{\Gamma(r+t)}{\Gamma(r)\Gamma(t)} \int_a^x \frac{(u-a)^{r-1}(b-u)^{t-1}}{(b-a)^{r+t-1}} du$	b	$\sigma = \frac{b-a}{r+t}\sqrt{\frac{rt}{r+t+1}}$
	$r > 0$	
	$t > 0$	

a brief introduction to the *central limit theorem* and the derived *Normal* and *Lognormal distributions* is given.

4.3.13 The Central Limit Theorem and Derived Distributions

The *central limit theorem* states:

The probability distribution for the sum of a number of random variables approaches the *Normal distribution* as the number becomes large.

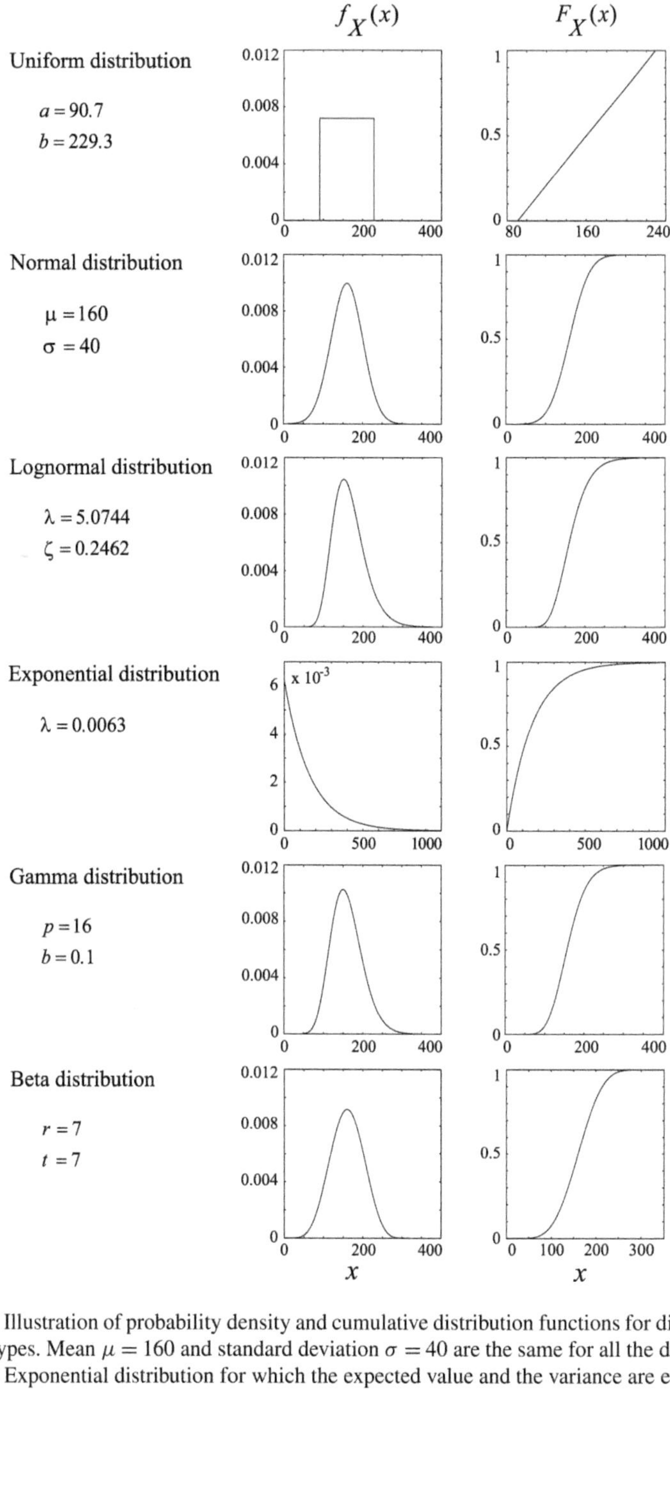

Fig. 4.11 Illustration of probability density and cumulative distribution functions for different distribution types. Mean $\mu = 160$ and standard deviation $\sigma = 40$ are the same for all the distributions except the Exponential distribution for which the expected value and the variance are equal

This result, which indeed is one of the most important results in probability theory, will not be derived here but instead the general conditions for the validity of the theorem will be outlined.

In principle, the theorem is valid as long as the number of independent contributions to the sum is "large". This implies that the sum may not be dominated by one or just a few random variables and furthermore, that the dependency between the random variables in the sum is not too strong. There is no requirement on the type of distributions of the random variables entering the sum, but if the distributions are skewed the number of variables in the sum which is required for the validity of the theorem increases.

4.3.14 Example 4.5—Central Limit Theorem

For the purpose of illustration, consider the problem of assessing the accumulated error in repeated measurements. The length of a structural member is being measured using a ruler of length 2 m with the smallest measuring unit equal to 1 mm. It is assumed that all measurements are rounded off to the closest unit on the ruler and thus it is assumed that each measurement is subject to a measurement uncertainty which is uniformly distributed in the range ± 0.5 mm. If the length of a considered structural member is smaller or equal to 2 m, the length can be measured by one measurement. It is clear that in this case the measurement uncertainty follows the uniform distribution as outlined in the above. However, if the member length is between 2 m and 4 m two measurements are required, if the member length is between 4 m and 8 m three measurements are required and so on. In Fig. 4.12, the histograms of the corresponding resulting measurement errors are illustrated under the assumption that consecutive errors are independent.

From Fig. 4.12, it is seen that whereas the sample histogram for one measurement is clearly uniform, the histogram approaches a bell shape already for four repeated measurements and for most practical purposes may be considered to be Normal distributed already for eight repeated measurements. The analytical form of the probability density function for the accumulated errors may be derived by repeated use of the result concerning the probability density function for the sum of random variables given in Eq. 4.34. In Benjamin and Cornell [4], it is heuristically shown that the analytical probability density functions has the form of a Normal distribution.

4.3.15 The Normal Distribution

The significant practical importance of the central limit theorem lies in the fact that even though only weak information is available regarding the number of contributions and their joint probability density function rather strong information is achieved for the distribution of sum of the contributions.

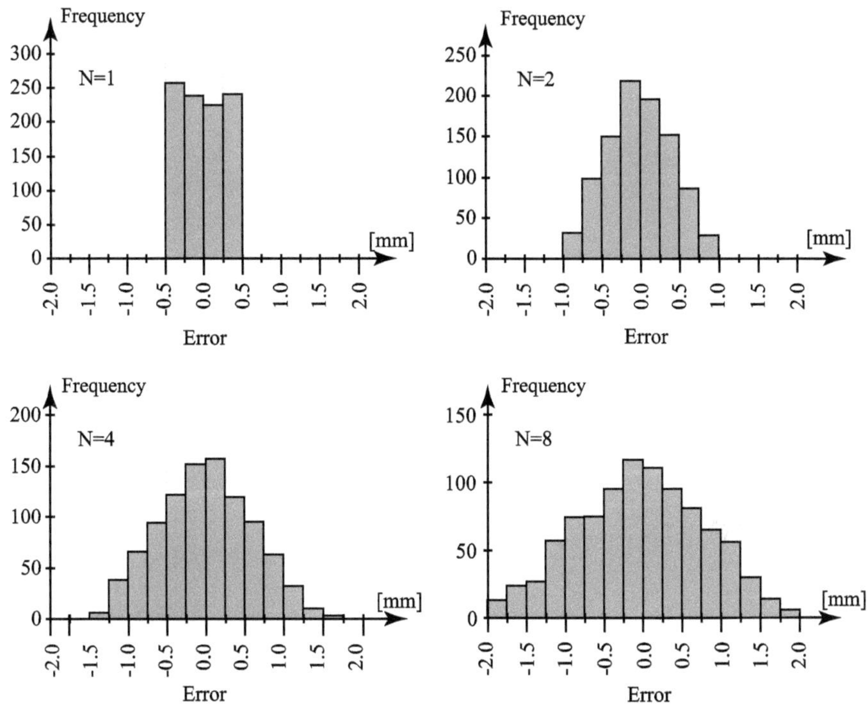

Fig. 4.12 Sample histograms for errors accumulated in 1, 2, 4 and 8 repeated measurements

The *Normal probability distribution* is thus applied very frequently in practical problems for the probabilistic modeling of uncertain phenomena which may be considered to originate from a cumulative effect of several independent uncertain contributions.

The Normal distribution has the property that the linear combination S of n Normal distributed random variables X_i, $i = 1, 2, \ldots, n$:

$$S = a_0 + \sum_{i=1}^{n} a_i X_i \tag{4.48}$$

is also Normal distributed. The distribution is said to be closed in respect to summation.

Figure 4.13 shows the probability density and distribution function of a Normal distributed random variable. The area between the interval of $\mu - \sigma \leq x \leq \mu + \sigma$ for the density function corresponds to 68.3% of the total area of the density function which always has the value 1.

One special version of the Normal distribution should be mentioned, namely the *Standard Normal distribution*. In general a standardized (sometimes referred to as a reduced) random variable is a random variable which has been transformed such that it has an expected value $E[X] = 0$ and a variance $V[X] = 1$, i.e. the random

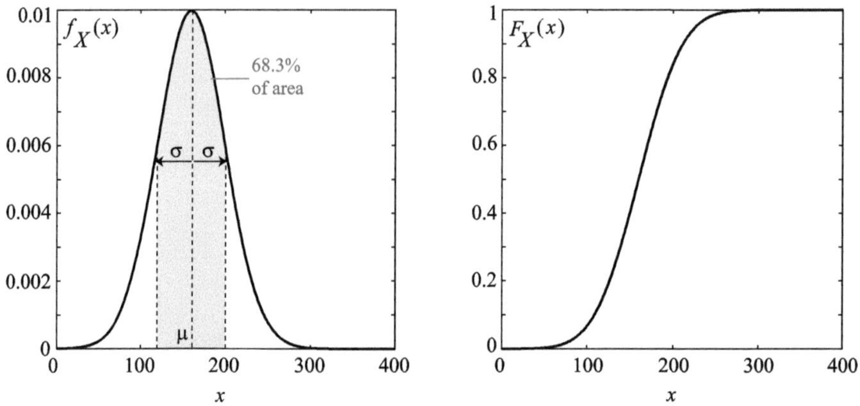

Fig. 4.13 Example of the density and distribution function of a normally distributed random variable defined by the parameters $\mu = 160$ and $\sigma = 40$

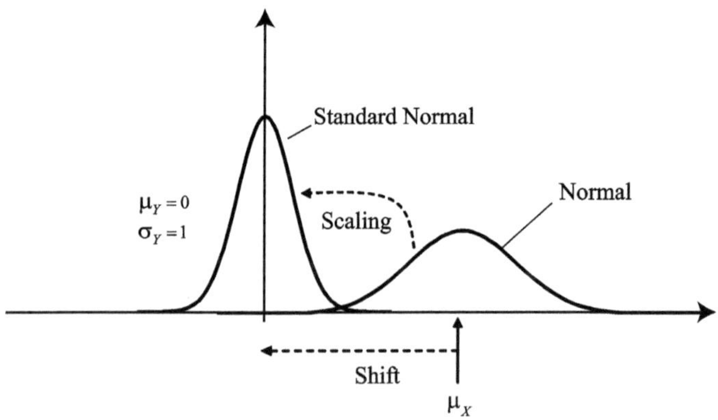

Fig. 4.14 Illustration of the relationship between a Normal distributed random variable and a standard Normal distributed random variable

variable Y defined by:

$$Y = \frac{X - \mu_X}{\sigma_X} \tag{4.49}$$

is a standardized random variable. If the random variable X follows the Normal distribution the random variable Y is standard Normal distributed. In Fig. 4.14 the process of standardization is illustrated.

It is common practice to denote the cumulative distribution function for the standard Normal distribution by $\Phi(x)$ and the corresponding density function by $\varphi(x)$. These functions are broadly available in software packages such as MS Excel and Matlab.

4.3.16 The Lognormal Distribution

A random variable Y is said to be *Lognormal distributed* if the variable $Z = \ln(Y)$ is Normal distributed. It thus follows that if an uncertain phenomenon can be assumed to originate from a multiplicative effect of several uncertain contributions, then the probability distribution for the phenomenon can be assumed to be Lognormal distributed.

The Lognormal distribution has the property that if:

$$P = \prod_{i=1}^{n} Y_i^{a_i} \tag{4.50}$$

and all Y_i are independent Lognormal random variables with parameters λ_i, ζ_i and $\varepsilon_i = 0$ as given in Table 4.2 then also P is Lognormal with parameters:

$$\lambda_P = \sum_{i=1}^{n} a_i \lambda_i \tag{4.51}$$

$$\zeta_P^2 = \sum_{i=1}^{n} a_i^2 \zeta_i^2 \tag{4.52}$$

4.4 Stochastic Processes and Extremes

Random quantities may be "*time variant*" in the sense that they take on new realizations at new trials or at new times. If the new realizations of the time variant random quantity occur at discrete times and take on discrete realizations, the random quantity is usually denoted a *random sequence*. Well known examples are series of throws of dice; more engineering relevant examples are flooding events. If the realizations of the time variant quantity occur continuously in time and take on continuous realizations the random quantity is usually denoted a *random process* or *stochastic process*. Examples include wind velocity, wave heights, snowfall and water levels.

In some cases random sequences and random processes may be represented in a given problem context in terms of random variables e.g. for the modeling of the "*point in time*" value of the intensity of the wind velocity, or the maximum (extreme) wind velocity during one year. However, in many cases this is not possible and then it is necessary to model the uncertain phenomena by a random process. In the following, first an important type of random sequence will be introduced, namely the sequence of *Bernoulli trials* from which the *Binomial distribution* has been derived. Thereafter, a description of the *Poisson counting process* is given and finally the continuous Normal or *Gaussian processes* are described. It should be noted that numerous other types of random processes have been suggested in the literature of which most have been derived from the mentioned.

4.4.1 Random Sequences—Bernoulli Trials

A sequence of experiments with only two possible mutually exclusive outcomes is called a sequence of Bernoulli trials. Typically the two possible events of a Bernoulli trial are referred to as a *success* or a *failure*. If it is assumed that the probability of success of a Bernoulli trial is constant equal to p then the probability density of Y successes in n trials $p_Y(y)$ i.e. the *Binomial distribution* (or sometimes denoted $B(n,p)$) can be shown to be equal to:

$$p_Y(y) = \binom{n}{y} p^y (1-p)^{n-y}, \quad y = 0, 1, 2, \ldots, n \tag{4.53}$$

where $\binom{n}{y}$ is the *binomial operator* defined as:

$$\binom{n}{y} = \frac{n!}{y!(n-y)!} \tag{4.54}$$

The cumulative distribution function for Y is thus given as:

$$P(Y \le y) = P_Y(y) = \sum_{i=0}^{y} \binom{n}{i} p^i (1-p)^{n-i}, \quad y = 0, 1, 2, \ldots, n \tag{4.55}$$

In Fig. 4.15 some examples of the *Binomial distribution* are shown for $n = 5$.

The expected value and the variance of Y, i.e. $E[Y]$and $Var[Y]$ can be shown to be given as:

$$EY = np \tag{4.56}$$

$$Var[Y] = np(1-p) \tag{4.57}$$

It is often of significant interest to assess the statistical characteristics of the random "time" or random number of trials n until the first success occurs. The probability density of this event, provided that the trials are independent, is given by the so-called *Geometric distribution*:

$$p_N(n) = p(1-p)^{n-1} \tag{4.58}$$

and the corresponding cumulative distribution function by:

$$P_N(n) = \sum_{i=1}^{n} p(1-p)^{i-1} = 1 - (1-p)^n \tag{4.59}$$

The mean value and the variance of the *Geometric distribution* are given by:

$$E[N] = \frac{1}{p} \tag{4.60}$$

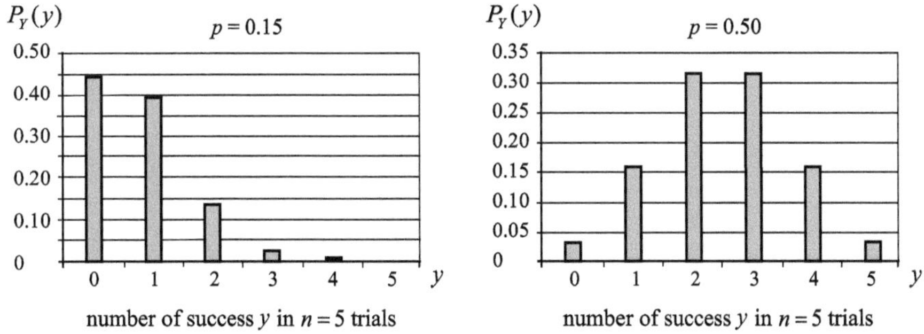

Fig. 4.15 Binomial distribution for $p = 0.15$ and $p = 0.50$, respectively

$$Var[N] = \frac{1-p}{p^2} \tag{4.61}$$

Especially the result given in Eq. 4.60 is of practical value as it gives the average "time" until success.

4.4.2 Example 4.6—Quality Control of Concrete

During the control procedures for the concrete quality on a building site not all of the tested specimens fulfilled the requirements to be used for structural purposes. The "rate of success" is the probability p that one specimen does not fulfil the requirements. According to the recent measurements $p = 0.2$. If we want to know after how many tests one specimen may fail the requirements we could assess the average return period which corresponds to the estimation operator described in Eq. 4.60.

$$E[N] = \frac{1}{p} = \frac{1}{0.2} = 5$$

Lecture 7 (Aim of the Present Lecture) The aim of the present lecture is to provide an understanding on how to model and probabilistically describe events that occur at discrete time points. In addition, continuous random processes are introduced and their main characteristics are provided. Finally, extreme events and their modeling are introduced and the concept of return period is explained.
On the basis of the lecture it is expected that the reader will acquire knowledge and skills with regard to:

- What is a simple Poisson process and where can it be applied?
- What are the properties which must be fulfilled before we can assume a Poisson process?

- What does homogeneity refer to for a Poisson process?
- According to what distribution can the time between realizations of a Poisson process be modeled?
- Which distribution does the sum of independent exponentially distributed variables follow and for what can this distribution be applied?
- What is a continuous random process?
- How can a Normal process be characterized?
- What does stationarity mean and how is it defined?
- What is an extreme value and what is required to model it probabilistically?
- Which are the different types of extreme value distributions?
- How are extreme value models and return periods related?

4.4.3 The Poisson Counting Process

The most commonly applied family of discrete processes in structural reliability are the *Poisson processes*. Due to the fact that Poisson processes have found applications in many different types of engineering problems, a large number of different variants of Poisson processes has evolved. In general, the process $N(t)$ denoting the number of points in the interval $[0; t[$is called a *simple Poisson process* if it satisfies the following conditions:

- The probability of one event in the interval $[t; t + \Delta t[$is asymptotically proportional to the length of the interval Δt.
- The probability of more than one event in the interval $[t; t + \Delta t[$ is a function of a higher order term of Δt for $\Delta t \to 0$.
- Events in disjoint intervals are mutually independent.

The Poisson process may be defined completely by its intensity $\nu(t)$:

$$\nu(t) = \lim_{\Delta t \to 0} \frac{1}{\Delta t} P \quad (\text{one event in } [t; t + \Delta t[) \tag{4.62}$$

If $\nu(t)$ is constant in time the Poisson process is said to be *homogeneous*, otherwise it is *non-homogeneous*.

In general, the probability of n events in the interval $[0; t[$ of a Poisson process with *intensity* $\nu(t)$ can be given as:

$$P_n(t) = \frac{\left(\int_0^t \nu(\tau)\, d\tau\right)^n}{n!} \exp\left(-\int_0^t \nu(\tau)\, d\tau\right) \tag{4.63}$$

with mean value $E[N(t)]$ and variance $Var[N(t)]$:

$$E\left[N(t)\right] = Var\left[N(t)\right] = \int_0^t \nu(\tau)\, d\tau \tag{4.64}$$

The probability of no events in the interval $[0; t[$ i.e. $P_0(t)$ is especially interesting considering reliability problems. This probability may be determined directly from Eq. 4.63 as:

$$P_0(t) = \exp\left(-\int_0^t \nu(\tau)\,d\tau\right) \tag{4.65}$$

implying that the time until and between events is *Exponential distributed*.

From Eq. 4.65 the cumulative distribution function of the *waiting time* until the first event T_1, i.e. $F_{T_1}(t_1)$ may be straightforwardly derived. Recognizing that the probability of $T_1 > t$ is $P_0(t)$ there is:

$$F_{T_1}(t_1) = 1 - \exp\left(-\int_0^{t_1} \nu(\tau)\,d\tau\right) \tag{4.66}$$

Consider now the sum of n independent and Exponential distributed waiting times T given as:

$$T = T_1 + T_2 + \cdots + T_n \tag{4.67}$$

It can be shown by repeated application of the result on the probability distribution for the sum of two random variables (see Eq. 4.34) that T is *Gamma distributed*:

$$f_T(t) = \frac{\nu(\nu t)^{(n-1)}\exp(-\nu t)}{(n-1)!} \tag{4.68}$$

4.4.4 Continuous Random Processes

A *random process* $X(t)$ is a random function of time, meaning that for any point in time, the value of $X(t)$ is a random variable. A realization of a random process (e.g. water level variation) is illustrated in Fig. 4.16.

A *stochastic process* $X(t)$ is said to be Normal or equivalently Gaussian if any set of random variables $X(t_i)$, $i = 1, 2, \ldots, n$ is jointly Normal distributed.

In accordance with the definition of the mean value of a random variable the mean value of all the possible realizations of the stochastic process at time t is given by:

$$\mu_X(t) = \int_{-\infty}^{\infty} x f_X(x; t)\,dx \tag{4.69}$$

The correlation between all the possible realizations at two points in time t_1 and t_2 as illustrated in Fig. 4.17 is described through the *autocorrelation function* $R_{XX}(t_1, t_2)$. Auto means that the function refers to only one realization.

The autocorrelation function is defined by:

$$R_{XX}(t_1, t_2) = E\left[X(t_1)X(t_2)\right] = \int_{-\infty}^{\infty}\int_{-\infty}^{\infty} x_1 x_2 f_{XX}(x_1, x_2; t_1, t_2)\,dx_1\,dx_2 \tag{4.70}$$

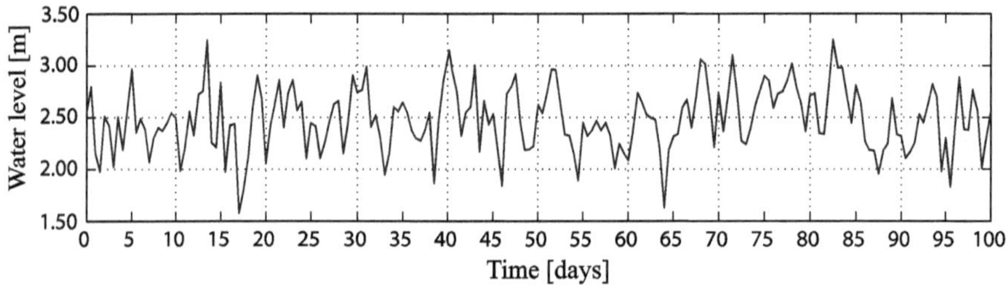

Fig. 4.16 Realization of the water level variation as function of time

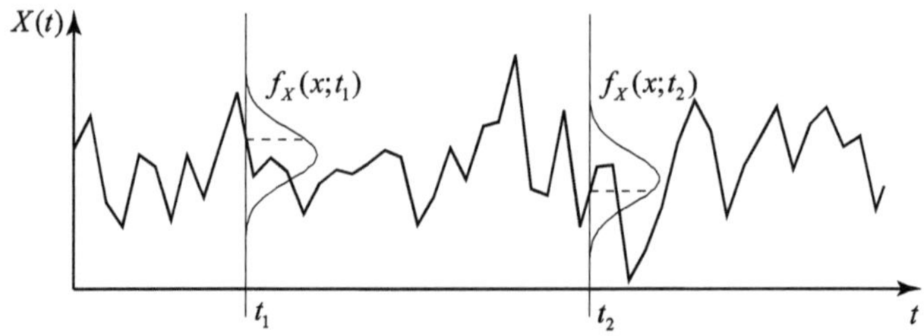

Fig. 4.17 Realization of a stochastic process with density functions of all possible realizations at two different points in time, t_1 and t_2

The auto-covariance function is defined as:

$$C_{XX}(t_1,t_2) = E\big[(X(t_1) - \mu_X(t_1))(X(t_2) - \mu_X(t_2))\big]$$
$$= \int_{-\infty}^{\infty}\int_{-\infty}^{\infty}(x_1 - \mu_X(t_1))(x_2 - \mu_X(t_2))\, f_{XX}(x_1,x_2;t_1,t_2)\,dx_1\,dx_2 \tag{4.71}$$

for $t_1 = t_2 = t$ the autocovariance function becomes the *covariance function*:

$$\sigma_X^2(t) = C_{XX}(t,t) = R_{XX}(t,t) - \mu_X^2(t) \tag{4.72}$$

where $\sigma_X(t)$ is the *standard deviation function*. The autocorrelation function can hence be expressed as:

$$R_{XX}(t_1,t_2) = \frac{C_{XX}(t_1,t_2)}{\sigma_X(t_1)\sigma_X(t_2)} \tag{4.73}$$

The above definitions for the scalar process $X(t)$ may be extended to cover also *vector valued processes* $\mathbf{X}(t) = (X_1(t), X_2(t), \ldots, X_n(t))^T$ having covariance functions $C_{X_iX_j} = Cov[X_i(t_1), X_j(t_2)]$. For $i = j$ these become the autocovariance functions and when $i \neq j$ these are termed as *cross-covariance functions*. An illustration of realizations of two stochastic processes with density functions of all possible realizations at two different points in time is shown in Fig. 4.18.

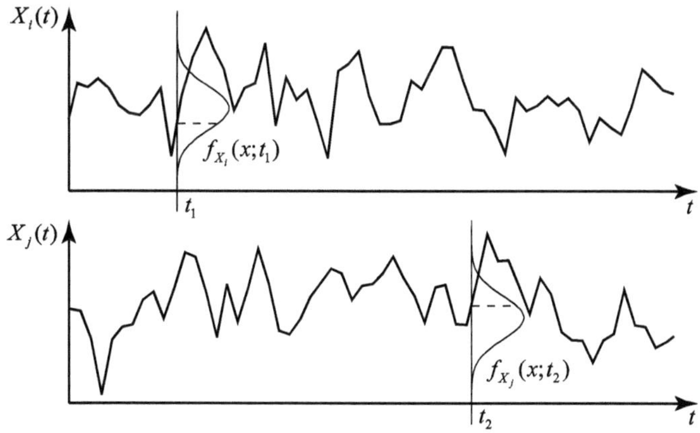

Fig. 4.18 Realizations of two stochastic processes with density functions of all possible realizations at two different points in time, t_1 and t_2

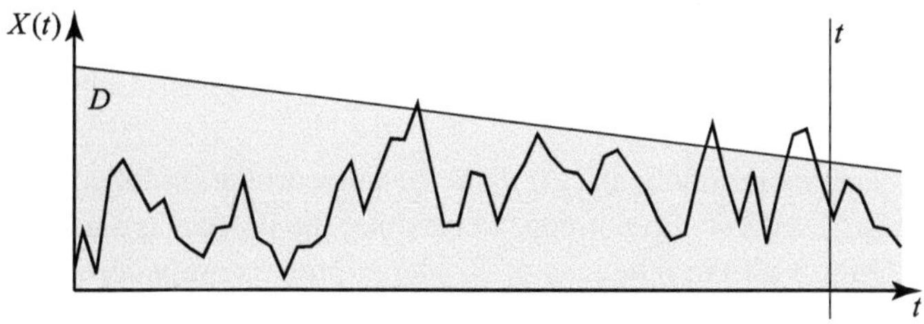

Fig. 4.19 Realizations of a stochastic process with three out-crossings out of the safe domain D in the time interval $[0; t]$

Finally the *correlation function* may be defined as:

$$\rho\left[X_i(t_1), X_j(t_2)\right] = \frac{Cov[X_i(t_1), X_j(t_2)]}{\sigma_{X_i}(t_1)\sigma_{X_j}(t_2)} \tag{4.74}$$

Typically the correlation function is an exponentially decaying function in time.

Having defined the mean value function and the autocovariance function for the stochastic process $X(t)$ the probability that the process remains within a certain safe domain D in the time interval $[0; t]$ may be evaluated by:

$$P_f(t) = 1 - P(N(t) = 0 | X(0) \in D) P(X(0) \in D) \tag{4.75}$$

where $N(t)$ is the number of out-crossings of the random process out of the domain D in the time interval $[0, t]$. This is illustrated in Fig. 4.19.

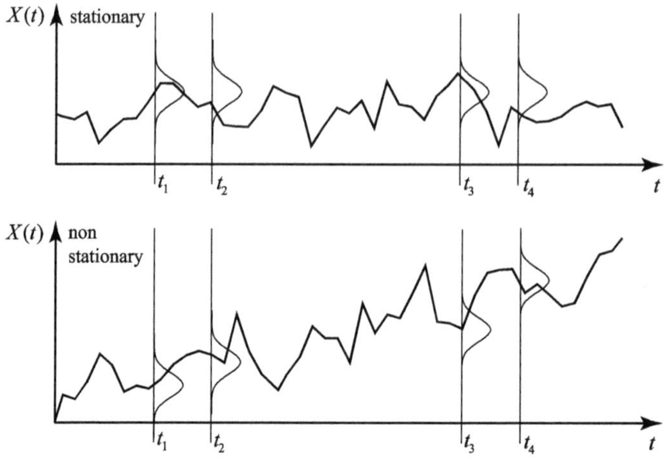

Fig. 4.20 Realizations of a stationary and a non stationary stochastic process

4.4.5 Stationarity and Ergodicity

When the mean value function $\mu_X(t)$ and the autocorrelation function $R_{XX}(t)$ of a stochastic process $X(t)$ do not depend on time, the process is said to be *weakly stationary*. Only if all the moments of a random process are independent of time, is the random process said to be *strictly stationary*. Illustrations of a stationary and non-stationary stochastic process are shown in Fig. 4.20.

A consequence of stationarity is that the autocovariance functions and autocorrelation function only depend on the time difference $\tau = t_1 - t_2$. In this case Eq. 4.70 may be written as:

$$R_{XX}(\tau) = E\left[X(t)X(t+\tau)\right] \tag{4.76}$$

It should be noted that for weakly stationary Normal stochastic processes, the requirements for strict stationarity are automatically fulfilled as the Normal distribution function is completely defined by the first two moments.

Stationarity, in principle, implies that the process cannot start or stop. However, for practical purposes, this requirement may be relaxed if the process is considered at a sufficient time after its start and/or before its end. Further, stationarity may be assumed even for slowly varying stochastic processes if sufficiently short time intervals are considered.

If, in addition to stationarity, the mean value function and the autocorrelation function of a stochastic process may be defined by a time average over one realization of the stochastic process, the process is said to be *weakly ergodic*. If all moments of the process may be defined in this way, the process is said to be *strictly ergodic*.

The assumption of *ergodicity* is especially important for the estimation of the statistical characteristics of stochastic processes when only one or a few realizations of the process are available, provided they are sufficiently long. If, for example, an engineer wants to model the water temperature at a coastline with a random variable, assuming ergodicity, she/he can use data from one single measurement station to estimate the distribution parameters as long as the data stream is long enough. In practice, ergodicity is, in such cases, often assumed unless, of course, evidence of physical understanding supports the contrary.

4.4.6 Statistical Assessment of Extreme Values

In risk and reliability assessments, *extreme values* (small and large) of random processes in a specified reference period are often of special interest. This is e.g. the case when considering the maximum sea water level, maximum wave heights, minimum ground water reservoir level, maximum wind pressures, strength of weakest link systems, maximum snow loads, etc.

For continuous time-varying loads, which can be described by a scalar, i.e. the water level or the wind pressure, one can define a number of related probability distributions. Often the simplest, namely the "*arbitrary point in time*", distribution is considered.

If $x(t^*)$ is a realization of a single time-varying load at time t^* then $F_X(x)$ is the arbitrary point in time cumulative distribution function of $X(t)$ defined by:

$$F_X(x) = P(X(t^*) \leq x) \tag{4.77}$$

In Fig. 4.21 first observations of half yearly maximum values of wind speeds are plotted together with histograms showing the corresponding sample frequency distributions. In the same figure the equivalent presentations are also provided for observations corresponding to maximums observed over periods of one and five years.

From Fig. 4.21 it is seen that there is a clear tendency that the mean value of the sample frequency histograms increases with increasing length of the considered period. At the same time, the standard deviation is seen to be decreasing.

For practical purposes, the observations of half yearly maxima may be assumed to be statistically independent and provide the basis (random "half yearly" point in time model) for the further modeling of the statistical characteristics of extremes for longer periods by extreme value considerations.

In the following section some results are given concerning the extreme events of trials of random variables and random processes, see also Madsen et al. [9] and Benjamin and Cornell [4]. Taking basis in the tail behavior of cumulative distribution functions, asymptotic results are given leading to the *extreme value distributions*.

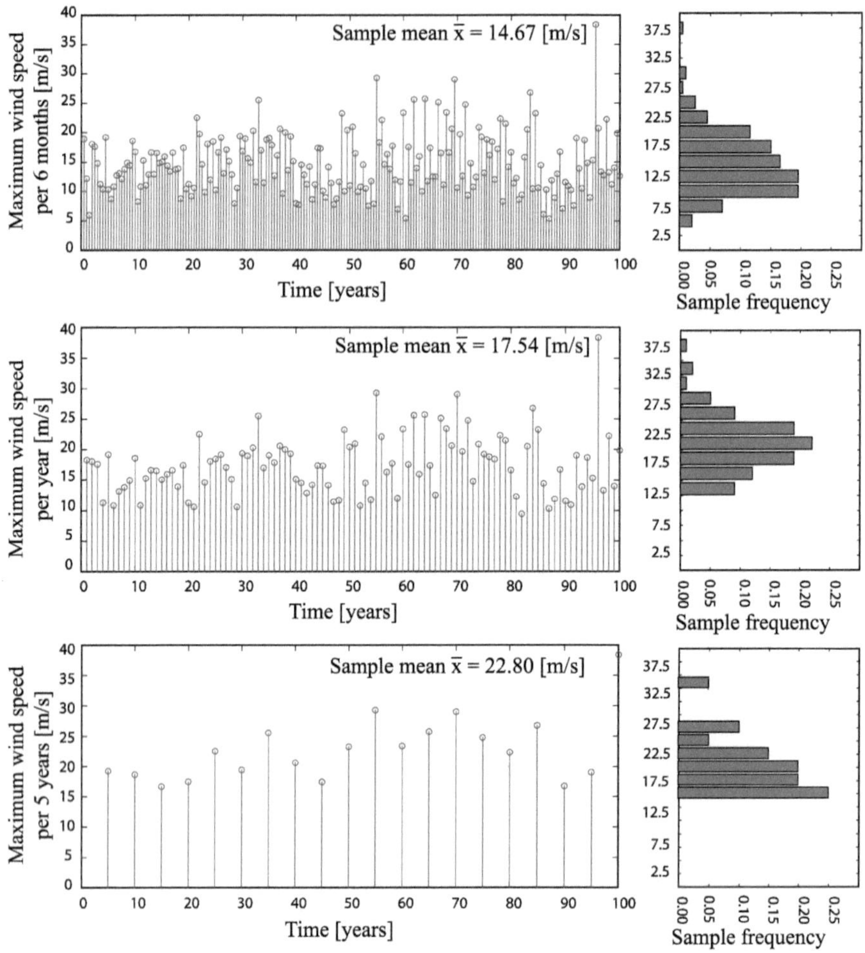

Fig. 4.21 Time series and corresponding sample frequency histograms of recorded half yearly, annual and five year maximum observed wind velocities

4.4.7 *Extreme Value Distributions*

When *extreme events* are of interest, the arbitrary point in time distribution of the load variable is not of immediate relevance but rather the distribution of the maximal values of the considered quantity over a given reference period.

If the random process $X(t)$ may be assumed to be ergodic, the distribution of the largest extreme in a *reference period* T, $F_{X,T}^{\max}(x)$ can be thought of as being generated by sampling values of the maximal realization $x_{\max}$ from successive reference periods T.

The cumulative distribution function of the largest extreme in a period of nT, $F_{X,nT}^{\max}(x)$, (with n being an integer) may be determined from the cumulative distri-

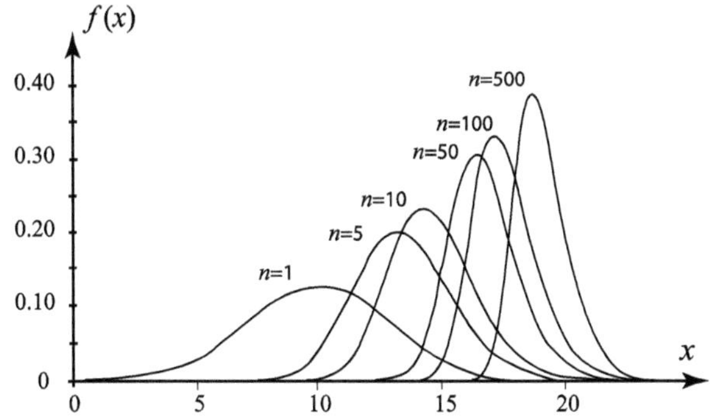

Fig. 4.22 Normal extreme value probability density functions

bution function of the largest extreme in the period T, $F_{X,T}^{\max}(x)$, by:

$$F_{X,nT}^{\max}(x) = \left[F_{X,T}^{\max}(x)\right]^n \tag{4.78}$$

which follows from the multiplication law for independent events. The corresponding probability density function may be established by differentiation of Eq. 4.78 yielding:

$$f_{X,nT}^{\max}(x) = nF_{X,T}^{\max}(x)^{n-1} f_{X,T}^{\max}(x) \tag{4.79}$$

In Fig. 4.22 the case of a Normal distribution with mean value equal to 10 and standard deviation equal to 3 is illustrated for increasing n.

Similar to the derivation of Eq. 4.78 the cumulative distribution function for the extreme minimum value in a considered reference period T, $F_{X,nT}^{\min}(x)$ may be found as:

$$F_{X,nT}^{\min}(x) = 1 - \left(1 - F_{X,T}^{\min}(x)\right)^n \tag{4.80}$$

Subject to the assumption that the considered process is ergodic, it can be shown that the cumulative function for an extreme event $F_{X,nT}^{\max}(x)$ converges asymptotically (as the reference period nT increases) to one of three types of extreme value distributions, type I, type II, or type III. To which type the distribution converges depends only on the tail behavior (upper or lower) of the considered random variable generating the extremes, i.e. $F_{X,T}^{\max}(x)$. In the following sections the three types or *extreme value distributions* will be introduced and it will be discussed under what conditions they may be assumed. In Table 4.2 the definition of the extreme value probability distributions and their parameters and moments is summarized.

4.4.8 Type I Extreme Maximum Value Distribution—Gumbel Max

For upwards unbounded distribution functions $F_X(x)$ where the upper tail falls off in an exponential manner, such as the case of the exponential function, the Normal

Table 4.2 Probability distributions, Schneider [11]

Distribution type	Parameter	Mean and standard deviation
Extrem Typ I—Gumbel max		
$-\infty \leq x \leq \infty$		
$f_X(x) = \alpha \exp(-\alpha(x-u) - \exp(-\alpha(x-u)))$	u	$\mu = u + \frac{0.577216}{\alpha}$
$F_X(x) = \exp(-\exp(-\alpha(x-u)))$	$\alpha > 0$	$\sigma = \frac{\pi}{\alpha\sqrt{6}}$
Extreme Type I—Gumbel min		
$-\infty \leq x \leq \infty$		
$f_X(x) = \alpha \exp(\alpha(x-u) - \exp(\alpha(x-u)))$	u	$\mu = u - \frac{0.577216}{\alpha}$
$F_X(x) = 1 - \exp(-\exp(\alpha(x-u)))$	$\alpha > 0$	$\sigma = \frac{\pi}{\alpha\sqrt{6}}$
Extreme Type II—Fréchet max		
$\varepsilon \leq x \leq \infty, u, k > 0$		
$f_X(x) = \frac{k}{u-\varepsilon}\left(\frac{u-\varepsilon}{x-\varepsilon}\right)^{k+1} \exp\left(-\left(\frac{u-\varepsilon}{x-\varepsilon}\right)^k\right)$	$u > 0$	$\mu = \varepsilon + (u-\varepsilon)\Gamma\left(1 - \frac{1}{k}\right)$
$F_X(x) = \exp\left(-\left(\frac{u-\varepsilon}{x-\varepsilon}\right)^k\right)$	$k > 0$	for $k > 1$
	ε	$\sigma = (u-\varepsilon)\sqrt{\Gamma\left(1 - \frac{2}{k}\right) - \Gamma^2\left(1 - \frac{1}{k}\right)}$
		for $k > 2$
Extreme Type III—Weibull min		
$\varepsilon \leq x \leq \infty, u, k > 0$		
$f_X(x) = \frac{k}{u-\varepsilon}\left(\frac{x-\varepsilon}{u-\varepsilon}\right)^{k-1} \exp\left(-\left(\frac{x-\varepsilon}{u-\varepsilon}\right)^k\right)$	$u > 0$	$\mu = \varepsilon + (u-\varepsilon)\Gamma\left(1 + \frac{1}{k}\right)$
$F_X(x) = 1 - \exp\left(-\left(\frac{x-\varepsilon}{u-\varepsilon}\right)^k\right)$	$k > 0$	$\sigma = (u-\varepsilon)\sqrt{\Gamma\left(1 + \frac{2}{k}\right) - \Gamma^2\left(1 + \frac{1}{k}\right)}$
	ε	$k \approx \left(\frac{\sigma}{\mu}\right)^{-1.09}$

distribution and the Gamma distribution, the cumulative distribution of extremes in the reference period T i.e. $F_{X,T}^{\max}(x)$ has the following form:

$$F_{X,T}^{\max}(x) = \exp\bigl(-\exp(-\alpha(x-u))\bigr) \tag{4.81}$$

with corresponding probability density function:

$$f_{X,T}^{\max}(x) = \alpha \exp\bigl(-\alpha(x-u) - \exp(-\alpha(x-u))\bigr) \tag{4.82}$$

which is also called the *Gumbel distribution* for extreme maxima. The mean value and the standard deviation of the Gumbel distribution may be related to the parameters u and α as:

$$\begin{aligned} \mu_{X_T^{\max}} &= u + \frac{\gamma}{\alpha} = u + \frac{0.577216}{\alpha} \\ \sigma_{X_T^{\max}} &= \frac{\pi}{\alpha\sqrt{6}} \end{aligned} \tag{4.83}$$

where γ is Euler's constant.

The Gumbel distribution has the useful property that the standard deviation is independent of the considered reference period, i.e. $\sigma_{X_{nT}^{\max}} = \sigma_{X_T^{\max}}$ and that the mean value $\mu_{X_{nT}^{\max}}$ depends on n in the following simple way:

$$\mu_{X_{nT}^{\max}} = \mu_{X_T^{\max}} + \frac{\sqrt{6}}{\pi}\sigma_{X_T^{\max}} \ln(n) \tag{4.84}$$

4.4.9 *Type I Extreme Minimum Value Distribution—Gumbel Min*

In case that the cumulative distribution function $F_X(x)$ is downwards unbounded and the lower tail falls off in an exponential manner, symmetry considerations lead to a cumulative distribution function for the extreme minimum $F_{X,T}^{\min}(x)$ within the reference period T of the following form:

$$F_{X,T}^{\min}(x) = 1 - \exp\big(-\exp(\alpha(x-u))\big) \tag{4.85}$$

with corresponding probability density function:

$$f_{X,T}^{\min}(x) = \alpha \exp\big(\alpha(x-u) - \exp(\alpha(x-u))\big) \tag{4.86}$$

which is also called the *Gumbel distribution* for extreme minima. The mean value and the standard deviation of the Gumbel distribution can be related to the parameters u and α as:

$$\begin{aligned} \mu_{X_T^{\min}} &= u - \frac{\gamma}{\alpha} = u - \frac{0.577216}{\alpha} \\ \sigma_{X_T^{\min}} &= \frac{\pi}{\alpha\sqrt{6}} \end{aligned} \tag{4.87}$$

4.4.10 *Type II Extreme Maximum Value Distribution—Fréchet Max*

For cumulative distribution functions downwards limited at zero and upwards unlimited with a tail falling off in the form:

$$F_X(x) = 1 - \beta\left(\frac{1}{x}\right)^k \tag{4.88}$$

the cumulative distribution function of extreme maxima in the reference period T i.e. $F_{X,T}^{\max}(x)$ has the following form:

$$F_{X,T}^{\max}(x) = \exp\left(-\left(\frac{u}{x}\right)^k\right) \tag{4.89}$$

with corresponding probability density function:

$$f_{X,T}^{\max}(x) = \frac{k}{u}\left(\frac{u}{x}\right)^{k+1} \exp\left(-\left(\frac{u}{x}\right)^{k}\right) \tag{4.90}$$

which is also called the *Fréchet distribution* for extreme maxima. The mean value and the variance of the Fréchet distribution can be related to the parameters u and k as:

$$\begin{aligned} \mu_{X_T^{\max}} &= u\Gamma\left(1-\frac{1}{k}\right) \\ \sigma_{X_T^{\max}}^2 &= u^2\left[\Gamma\left(1-\frac{2}{k}\right) - \Gamma^2\left(1-\frac{1}{k}\right)\right] \end{aligned} \tag{4.91}$$

where it is noticed that the mean value only exists for $k > 1$ and the standard deviation only exist for $k > 2$. In general it can be shown that the ith moment of the Fréchet distribution exists only when $k > i$.

4.4.11 Type III Extreme Minimum Value Distribution—Weibull Min

Finally, in the case where the cumulative distribution function $F_X(x)$ is downwards limited at ε and the lower tail falls off towards ε in the form:

$$F(x) = c(x-\varepsilon)^k \tag{4.92}$$

leads to a cumulative distribution function for the extreme minimum $F_{X,T}^{\min}(x)$ within the reference period T of the following form:

$$F_{X,T}^{\min}(x) = 1 - \exp\left(-\left(\frac{x-\varepsilon}{u-\varepsilon}\right)^{k}\right) \tag{4.93}$$

with corresponding probability density function:

$$f_{X,T}^{\min}(x) = \frac{k}{u-\varepsilon}\left(\frac{x-\varepsilon}{u-\varepsilon}\right)^{k-1} \exp\left(-\left(\frac{x-\varepsilon}{u-\varepsilon}\right)^{k}\right) \tag{4.94}$$

which is also called the *Weibull distribution* for extreme minima. The mean value and the variance of the Weibull distribution can be related to the parameters u, k and ε as:

$$\begin{aligned} \mu_{X_T^{\min}} &= \varepsilon + (u-\varepsilon)\Gamma\left(1+\frac{1}{k}\right) \\ \sigma_{X_T^{\min}}^2 &= (u-\varepsilon)^2\left[\Gamma\left(1+\frac{2}{k}\right) - \Gamma^2\left(1+\frac{1}{k}\right)\right] \end{aligned} \tag{4.95}$$

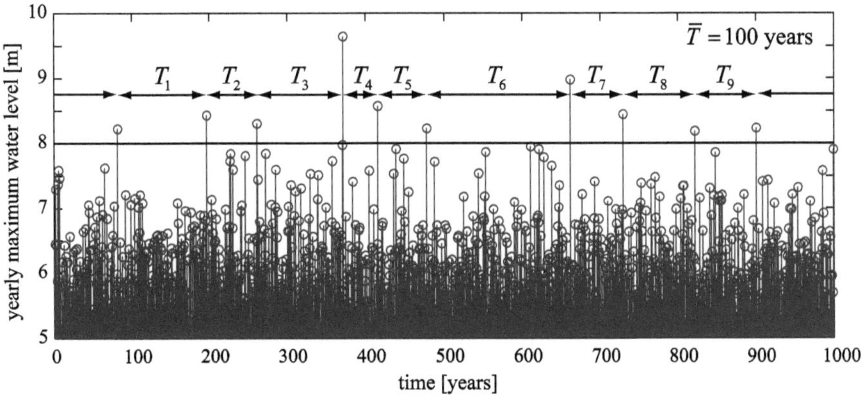

Fig. 4.23 Maximum water levels with threshold for the 100-year return period flood event

4.4.12 Return Period for Extreme Events

The *return period* T_R for an extreme event corresponding to x may be defined by:

$$T_R = nT = \frac{1}{(1 - F_{X,T}^{\max}(x))} \quad T \tag{4.96}$$

where T is the reference period for the cumulative distribution function of the extreme events $F_{X,T}^{\max}(x)$. If, for example, the annual probability of an extreme load event is 0.02, the return period for this load event is 50 years.

4.4.13 Example 4.7—A Flood with a 100-Year Return Period

The concept of the return period is often applied in the area of flood protection. In Fig. 4.23, the 100-year return period flood event is defined by the exceedance of a water level of 8 m. The yearly maximum water level is higher than this threshold on average every 100 years, although waiting times between two consecutive events can differ from this average.

The threshold defining an event with return period T_R is called the characteristic value x_c. This value has a probability of exceedance of $p = 1/T_R$ and can be determined as a quantile value of the corresponding extreme value distribution.

For the Gumbel max distribution, it is possible to calculate the *characteristic value* approximately. By manipulation of Eq. 4.81 it can be shown, by utilizing a Taylor expansion to the first order of $\ln(p)$ in $p = 1$, that the characteristic value x_c corresponding to an annual exceedance probability of p and corresponding return period $T_R = 1/p$ for a Gumbel max distribution for large return periods can be

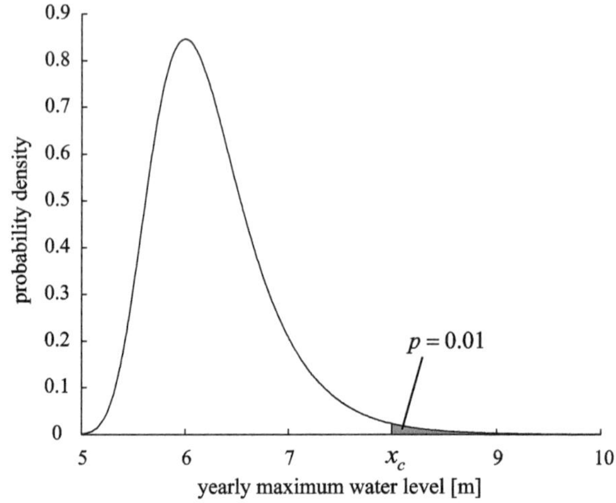

Fig. 4.24 Extreme value distribution of the yearly maximum water level with characteristic value for the 100-year return period flood event

written as:

$$x_c \approx u + \frac{1}{\alpha}\ln(T_R) \tag{4.97}$$

which shows that the characteristic value increases with the logarithm of the considered return period.

For this example, using a Gumbel max distribution with parameters $u = 6$ and $\alpha = 2.3$ and a return period $T_R = 100$, a characteristic value $x_c = 8$ is obtained. The threshold value of 8 m defining the 100-year return period flood event is shown in Figs. 4.23 and 4.24. While Fig. 4.24 shows the density function of the yearly maximum water level using a Gumbel max distribution with parameters as specified above, Fig. 4.23 has been obtained based on a simulation of the Gumbel max distribution with the same parameters.

4.5 Self Assessment Questions/Exercises

1. What types of uncertainties can be distinguished and how do these depend on the time and scale of modeling?
2. What is understood by the terms aleatory and epistemic uncertainties?
3. What is meant by the term "continuous random variable"?
4. Using the properties of the expectation operator, how may the following notations be rewritten? (Note that a and b are constants and X is a random variable)
   a. $E[a + bX]$
   b. $Var[a + bX]$
5. Write down the names of the axes of the probability density and the cumulative distribution functions of the random variable X illustrated in Figs. 4.25

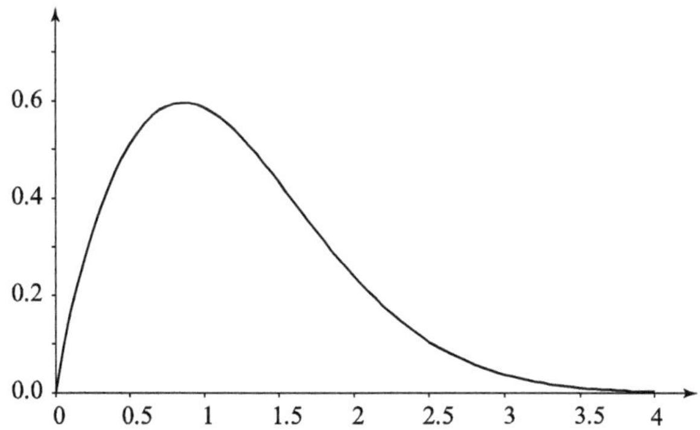

Fig. 4.25 Illustration of a probability density function

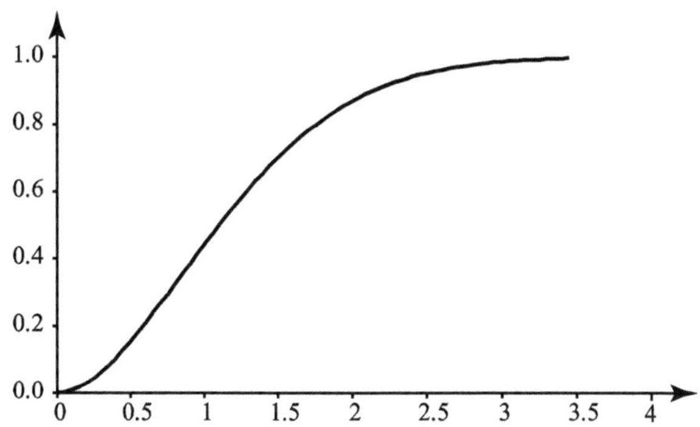

Fig. 4.26 Illustration of a cumulative distribution function

and 4.26. Identify the locations of the mean, the mode and the median in the illustration of the probability density function. Show also the value of the median in the illustration of the cumulative distribution function shown in Fig. 4.26.

6. State the central limit theorem.
7. What is a standardized random variable and how is it defined?
8. What is a Bernoulli trial and what does it describe?
9. What is a Poisson process and where can it be applied?
10. The probability density function of a continuous random variable X, defined in the interval $[0, 10]$, is illustrated in Fig. 4.27. Calculate the probability that X may exceed the value of 5.
11. It is given that the operational life (until breakdown) T of a diesel engine follows an exponential distribution, $F_T(t) = 1 - e^{-\lambda t}$, with parameter λ and mean value, $\mu_T = 1/\lambda$, equal to 10 years. Calculate the probability that the engine breaks down within 2 years after placed in operation.

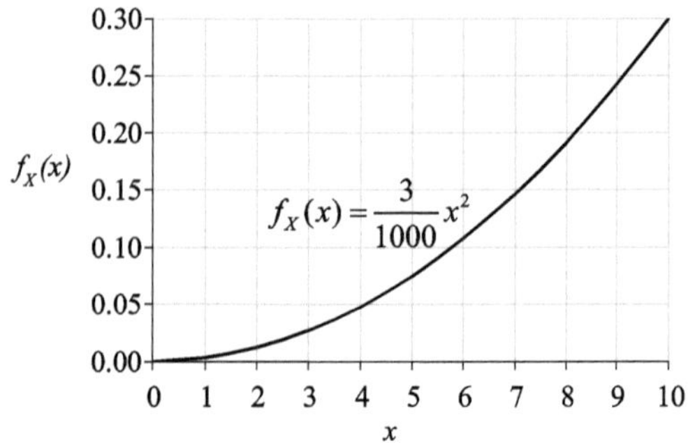

Fig. 4.27 Function of a continuous random variable X

12. In a city there are on average 5 snowfall events in a year. Assume that the occurrence of snowfall events follows a Poisson process. The number of snowfall events in t years, X, is described by the discrete cumulative distribution function $P(X = k) = \frac{(\nu t)^k}{k!} e^{-\nu t}$ with annual mean rate ν. How large is the probability of no snowfall in the next year? How large is the probability of exactly 5 snowfall events in the next year?

Chapter 5
Estimation and Model Building

Lecture 8 (Aim of the Present Lecture) The aim of the present lecture is to provide an overview of how to establish probabilistic models and to introduce the basic tools for assessing the validity of model assumptions. First, the problem of selecting appropriate probability distribution functions for the purpose of modeling uncertainties with basis in observations of uncertain phenomena is treated, and it is shown how the concept of probability paper can provide a pragmatic basis for this purpose. Then, the required theory and methodology for the estimation of parameters of probability distributions based on data is introduced. To this end, the method of moments and the maximum likelihood method as well as their limitations and applications are illustrated and discussed.
On the basis of the lecture it is expected that the reader will acquire knowledge and skills with regard to:

- What are the steps involved in establishing a probabilistic model for a random variable?
- What is a probability paper and how is it constructed? How is it related to a quantile plot?
- In what regions of the probability paper is it especially important that the plotted quantiles fit a straight line?
- What is the principle behind the method of moments and how can it be applied to estimate the parameters of a probability distribution?
- What is the principle behind the maximum likelihood method and how can it be applied to estimate the parameters of a probability distribution?
- What is sample likelihood and how is it quantified?
- How can the statistical uncertainty associated with estimated distribution parameters be quantified?
- What is the information matrix and how does this relate to the covariance matrix of the estimated parameters?

M.H. Faber, *Statistics and Probability Theory*,
Topics in Safety, Risk, Reliability and Quality 18,
DOI 10.1007/978-94-007-4056-3_5, © Springer Science+Business Media B.V. 2012

5.1 Introduction

An important task in risk and reliability analysis is to establish probabilistic models for the further statistical treatment of uncertain variables.

In the literature, a large number of probabilistic models for load and resistance variables may be found. For example, in the Probabilistic Model Code developed by the Joint Committee on Structural Safety (JCSS [8]) where probabilistic models may be found for the description of the strength and stiffness characteristics of steel and concrete materials, soil characteristics and for the description of load and load effects covering many engineering application areas. However, it is not always the case that an appropriate probabilistic model for the considered problem is available. Moreover, in engineering fields, such as in environmental engineering and hydrology standardization of the probabilistic modeling is relatively less developed. In such situations it is necessary that methodologies and tools are readily available for the statistical assessment of *frequentistic information* (e.g. observations and test results) and the formulation of *probabilistic models* of uncertain variables.

In practice two situations may thus be distinguished, namely the situation where a new probabilistic model is formulated from the very beginning and the situation where an already existing probabilistic model is updated on the basis of new information, e.g. observations or experimental results. The formulation of probabilistic models may be based on data (*frequentistic information*) alone, but most often data is not available to the extent where this is possible. In such cases, it is usually possible to base the model building on physical arguments, experience and judgment (*subjective information*). If also some data are available, the subjective information may be combined with the frequentistic information and the resulting probabilistic model is in effect of a Bayesian nature.

It should be emphasized that on the one hand the probabilistic model should aim for simplicity. On the other hand, the model should be accurate enough to allow for including important information collected during the lifetime of the considered technical system thereby facilitating the updating of the probabilistic model. In this way, uncertainty models, which initially are based entirely on subjective information will, as new information is collected, eventually be based on objective information.

In essence, the *model building* process consists of five steps:

- Assessment and statistical quantification of the available data
- Selection of distribution function
- Estimation of distribution parameters
- Model verification
- Model updating

Typically, the initial choice of the model, i.e. underlying assumptions regarding distributions and parameters may be based mainly on subjective information, whereas the assessment of the parameters of the distribution function and the verification of the models is performed on the basis of the available data. The principle for establishing a probabilistic model is illustrated in Fig. 5.1.

As the probabilistic models are based on both frequentistic information and subjective information, these are Bayesian in nature.

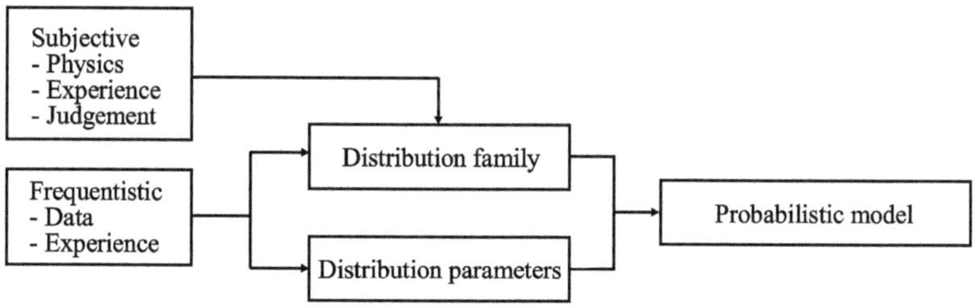

Fig. 5.1 Illustration of the formulation of probabilistic models for uncertain variables

In the following sections only the probabilistic modeling of random variables will be considered, but the described approach applies with some extensions also to the probabilistic modeling of random processes and random fields.

5.2 Selection of Probability Distributions

In general the distribution function for a given random variable or stochastic process is not known and must thus be chosen on the basis of frequentistic information, physical arguments or a combination of both.

A formal classical approach (described in detail in Sect. 5.9) for the identification of an appropriate distribution function on the basis of statistical evidence is to:

- Postulate a hypothesis for the distribution family.
- Estimate the parameters for the selected distribution on the basis of statistical data.
- Perform a statistical test to attempt to reject the hypothesis.

If it is not possible to reject the hypothesis, the selected distribution function may be considered to be appropriate for the modeling of the considered random variable. If the hypothesis is rejected, a new hypothesis must be postulated and the process repeated.

This procedure follows closely the classical frequentistic approach to statistical analysis. However, in many practical engineering applications this procedure has limited value. This is not only due to the fact that the amount of available data most often is too limited to form a solid basis for a statistical test, but also because the available tests applied in situations with little frequentistic information may lead to false conclusions.

In practice, however, it is often the case that physical arguments can be formulated for the choice of distribution functions and statistical data are therefore merely used for the purpose of checking whether the postulated distribution function is plausible.

A practically applicable approach for the selection of the distribution function for the modeling of a random variable is thus:

- First, to consider the physical reasons why the quantity at hand may belong to one or the other distribution family;
- Second, to check whether the statistical evidence is in gross contradiction with the assumed distribution by using e.g. probability paper as explained in the subsequent, or if relevant, the more formal approaches given in Sect. 5.9.

5.2.1 Model Selection by Use of Probability Paper

Having selected a probability distribution family for the probabilistic modeling of a random variable, the *probability paper* is an extremely useful tool for the purpose of checking the plausibility of the selected distribution family.

A probability paper for a given distribution family is constructed such that the cumulative probability distribution function (or the complement) for that distribution family will have the shape of a straight line when plotted on the paper. A probability paper is thus constructed by a non-linear transformation of the vertical axis (y-axis).

For a Normal distributed random variable the cumulative distribution function is given as:

$$F_X(x) = \Phi\left(\frac{x-\mu_X}{\sigma_X}\right) \tag{5.1}$$

where μ_X and σ_X are the mean value and the standard deviation of the Normal distributed random variable and where $\Phi(\cdot)$ is the standard Normal probability distribution function. Equation 5.1 can be re-written as:

$$x = \Phi^{-1}(F_X(x))\sigma_X + \mu_X \tag{5.2}$$

Now by plotting x against $\Phi^{-1}(F_X(x))$, see also Fig. 5.2, it is seen that a straight line is obtained with the slope depending on the standard deviation of the random variable X and the intercept on with the y-axis depending on the mean value of the random variable. Such a plot is sometimes called a quantile plot, see also Sect. 3.3.

Also in Fig. 5.2 the scale of the non-linear y-axis is given corresponding to the linear mapping of the observed cumulative probability densities. In probability papers, typically, only this non-linear scale is given.

Probability papers may also be constructed graphically. In Fig. 5.3 the graphical construction of a Normal probability paper is illustrated.

Various types of probability paper are readily available for use. Given an ordered set of observed values $\hat{\mathbf{x}}_i^o = (\hat{x}_i^o, \hat{x}_i^o, \ldots, \hat{x}_N^o)^T$ of a random variable, the cumulative distribution function may be evaluated as:

$$F_X(\hat{x}_i^o) = \frac{i}{N+1} \tag{5.3}$$

In Table 5.1 an example is given for a set of observed concrete cube compressive strength values together with the cumulative distribution function values as calculated using Eq. 5.3. In Fig. 5.4 the cumulative distribution values are plotted on a Normal distribution probability paper.

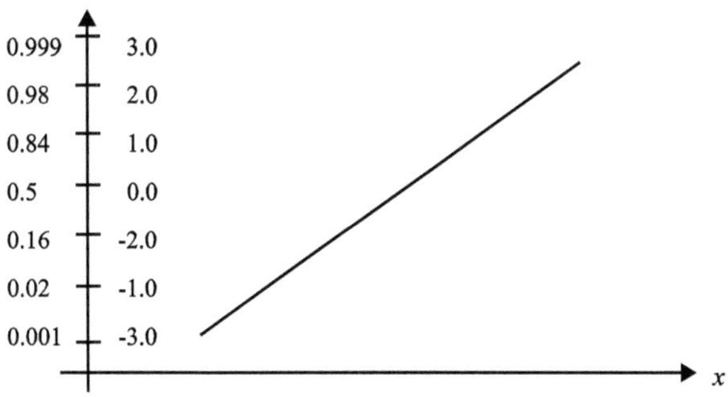

Fig. 5.2 Illustration of the non-linear scaling of the y-axis for a Normal distributed random variable

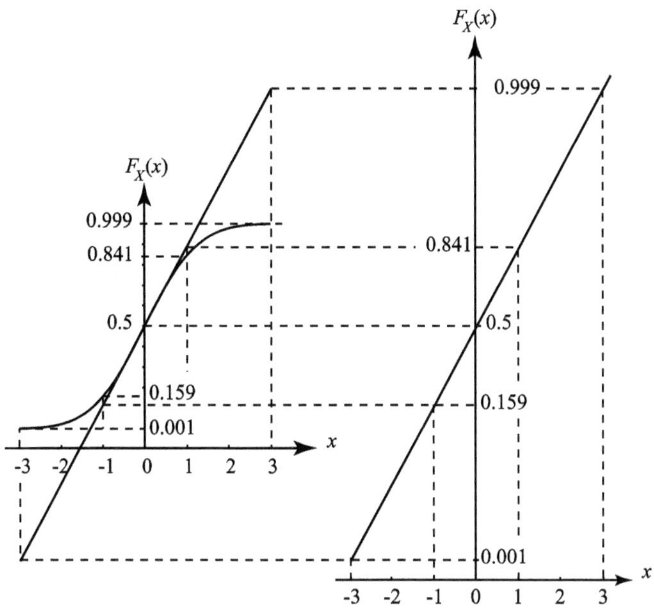

Fig. 5.3 Illustration of the graphical construction of a Normal distribution probability paper

A first estimate of the distribution parameters may be readily determined from the slope and the position of the best straight line through the plotted cumulative distribution values. In Sect. 5.3 the problem of parameter estimation is considered in more detail.

From Fig. 5.4 it is seen that the observed cumulative distribution function fits relatively well with a straight line. This might also be expected considering that the observed values of the concrete compressive strength are not really representative for the lower tail of the distribution, where due to the non-negativity of the com-

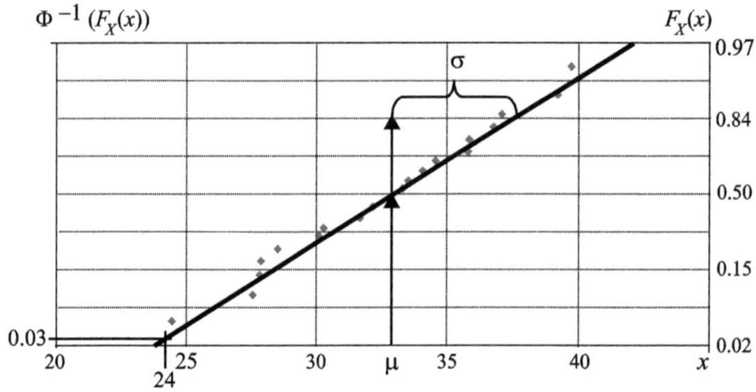

Fig. 5.4 Concrete cube compressive strength data plotted in a probability paper for the Normal distribution

Table 5.1 Ordered set of observed concrete cube compressive strengths and the calculated cumulative distribution values

i	$\hat{x}_i^o$	$F_X(\hat{x}_i^o)$	$\Phi(F_X(\hat{x}_i^o))$
1	24.4	0.048	−1.668
2	27.6	0.095	−1.309
3	27.8	0.143	−1.068
4	27.9	0.190	−0.876
5	28.5	0.238	−0.712
6	30.1	0.286	−0.566
7	30.3	0.333	−0.431
8	31.7	0.381	−0.303
9	32.2	0.429	−0.180
10	32.8	0.476	−0.060
11	33.3	0.524	0.060
12	33.5	0.571	0.180
13	34.1	0.619	0.303
14	34.6	0.667	0.431
15	35.8	0.714	0.566
16	35.9	0.762	0.712
17	36.8	0.810	0.876
18	37.1	0.857	1.068
19	39.2	0.905	1.309
20	39.7	0.952	1.668

pressive strength it might be assumed that a Lognormal distribution would be more suitable.

The estimation of the probabilities of values which are outside the measured range can be done by extrapolation (see Fig. 5.4). The probability paper may also be used for extreme phenomena such as the maximum water level, to estimate the values of the water level with a certain return period (see e.g. Schneider [11]). However, as always when extrapolating, extreme care must be exercised.

5.3 Estimation of Distribution Parameters

There are, in principle, two different methods to estimate the distribution parameters on the basis of data, namely the methods of *point estimates* and the methods of *interval estimates*. In the following section, however, only two of the methods of point estimates will be explained, namely the *method of moments* and the *method of maximum likelihood* as these have proven especially useful in practical risk and reliability engineering analysis.

5.3.1 The Method of Moments

Using the method of moments, the parameters of the distribution can be estimated by equating the moments obtained from the sample and the moments obtained from the distribution.

Assuming that the considered random variable X may be modeled by the probability density function $f_X(x; \boldsymbol{\Theta})$, where $\boldsymbol{\Theta} = (\theta_1, \theta_2, \ldots, \theta_k)^T$ are the distribution parameters, the first k moments $\lambda = (\lambda_1, \lambda_2, \ldots, \lambda_k)^T$ of the random variable X may be written as:

$$\begin{aligned} \lambda_j(\boldsymbol{\theta}) &= \int_{-\infty}^{\infty} x^j f_X(x|\boldsymbol{\theta})\, dx \\ &= \lambda_j(\theta_1, \theta_2, \ldots, \theta_k) \end{aligned} \tag{5.4}$$

If the random sample, from which the distribution parameters $\boldsymbol{\Theta} = (\theta_1, \theta_2, \ldots, \theta_k)^T$ are to be estimated, is collected in the vector $\hat{\mathbf{x}} = (\hat{x}_1, \hat{x}_{2,}, \ldots, \hat{x}_n)^T$, the corresponding k sample moments may be calculated as:

$$m_j = \frac{1}{n} \sum_{i=1}^{n} \hat{x}_i^j \tag{5.5}$$

By equating the k sample moments to the k moments of the random variable X, a set of k equations with the k unknown distribution parameters is obtained, the solution of which gives the *point estimates* of the distribution parameters.

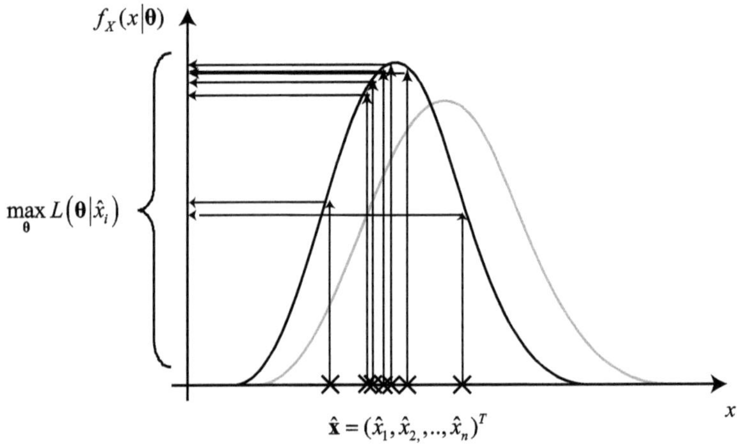

Fig. 5.5 Schematic representation of the method of maximum likelihood

5.3.2 *The Method of Maximum Likelihood*

This method may be somewhat more difficult to use than the method of moments but has a number of very attractive properties, which makes it especially applicable in engineering risk and reliability analysis.

The principle of the method is that the parameters of the distribution function are fitted such that the probability (likelihood) of the observed random sample is maximized.

Let the random variable of interest X have a probability density function $f_X(x;\boldsymbol{\theta})$ where $\boldsymbol{\theta} = (\theta_1, \theta_2, \ldots, \theta_k)^T$ are the distribution parameters to be estimated.

If the random sample from which the distribution parameters $\boldsymbol{\theta} = (\theta_1, \theta_2, \ldots, \theta_k)^T$ are to be estimated are collected in the vector $\hat{\mathbf{x}} = (\hat{x}_1, \hat{x}_{2,}, \ldots, \hat{x}_n)^T$, the likelihood $L(\boldsymbol{\Theta}|\hat{\mathbf{x}})$ of the observed random sample is defined as:

$$L(\boldsymbol{\theta}|\hat{\mathbf{x}}) = \prod_{i=1}^{n} f_X(\hat{x}_i|\boldsymbol{\theta}) \tag{5.6}$$

The maximum likelihood point estimates of the parameters $\boldsymbol{\theta} = (\theta_1, \theta_2, \ldots, \theta_k)^T$ may now be obtained by solving the following optimization problem:

$$\min_{\theta} (-L(\boldsymbol{\theta}|\hat{\mathbf{x}})) \tag{5.7}$$

The principle is illustrated in Fig. 5.5, where it can be seen that the distribution function is shifted to maximize the likelihood of the observations by means of changing the distribution parameters.

Instead of the likelihood function it is advantageous to consider the log-likelihood $l(\boldsymbol{\theta}|\hat{\mathbf{x}})$ i.e.:

$$l(\boldsymbol{\theta}|\hat{\mathbf{x}}) = \sum_{i=1}^{n} \ln(f_X(\hat{x}_i|\boldsymbol{\theta})) \tag{5.8}$$

One of the most attractive properties of the maximum likelihood method is that when the number of samples i.e. n is sufficiently large the distribution of the parameter estimates converges towards a Normal distribution with mean values $\boldsymbol{\mu}_{\boldsymbol{\Theta}}$ equal to the point estimates, i.e.:

$$\mu_{\boldsymbol{\Theta}} = (\theta_1^*, \theta_2^*, \ldots, \theta_n^*)^T \tag{5.9}$$

The covariance matrix $C_{\Theta\Theta}$ for the point estimates may readily be obtained by:

$$\mathbf{C}_{\boldsymbol{\Theta\Theta}} = \mathbf{H}^{-1} \tag{5.10}$$

where $\mathbf{H}$ is the *Fisher information matrix* with components determined by the second order partial derivatives of the log-likelihood function taken in the maximum, i.e.:

$$H_{ij} = -\left.\frac{\partial^2 l(\boldsymbol{\theta}|\hat{\mathbf{x}})}{\partial\theta_i \partial\theta_j}\right|_{\boldsymbol{\theta}=\boldsymbol{\theta}^*} \tag{5.11}$$

With the Fisher information matrix the uncertainties associated with the *point estimates* and the statistical uncertainties are considered (with a large number of samples, the uncertainties decrease).

5.3.3 Example 5.1—Parameter Estimation

Consider again the experimental results of the concrete cube compressive strength values given in Table 5.1. Assuming that the concrete cube compressive strength is Normal distributed, it is required now to estimate the parameters on the basis of the experiment results.

With Eqs. 4.5, 4.7 and 4.17 it can be shown that the first two moments of the Normal distribution are given as:

$$\begin{aligned} \lambda_1 &= \mu \\ \lambda_2 &= \mu^2 + \sigma^2 \end{aligned} \tag{5.12}$$

Analyzing the sample data, the first two sample moments can be found using the following equation:

$$\begin{aligned} m_j &= \frac{1}{n}\sum_{i=1}^{n} \hat{x}_i^j \\ m_1 &= \frac{1}{n}\sum_{i=1}^{n} \hat{x}_i = 32.67 \\ m_2 &= \frac{1}{n}\sum_{i=1}^{n} \hat{x}_i^2 = 1083.36 \end{aligned} \tag{5.13}$$

The point estimates of the parameters μ and σ may now be determined by solving the equations:

$$\begin{aligned} \mu &= 32.67 \\ \mu^2 + \sigma^2 &= 1083.36 \end{aligned} \tag{5.14}$$

giving:

$$\begin{aligned} \mu &= 32.67 \\ \sigma &= 4.05 \end{aligned} \tag{5.15}$$

Using the method of maximum likelihood, the maximum likelihood function is readily written as:

$$L(\boldsymbol{\theta}|\hat{\mathbf{x}}) = \prod_{i=1}^{n} f_X(\hat{x}_i|\boldsymbol{\theta}) = \left(\frac{1}{\sqrt{2\pi}\theta_2}\right)^n \exp\left(-\frac{1}{2}\sum_{i=1}^{n}\frac{(\hat{x}_i-\theta_1)^2}{\theta_2^2}\right) \tag{5.16}$$

and correspondingly the log-likelihood function as:

$$l(\boldsymbol{\Theta}|\hat{\mathbf{x}}) = n\ln\left(\frac{1}{\sqrt{2\pi}\theta_2}\right) - \frac{1}{2}\sum_{i=1}^{n}\frac{(\hat{x}_i-\theta_1)^2}{\theta_2^2} \tag{5.17}$$

where n is the number of observations of the concrete cube compressive strength $\hat{\mathbf{x}} = (\hat{x}_1, \hat{x}_{2,}, \ldots, \hat{x}_n)^T$.

The mean values of the estimates may be determined by solving the following equations:

$$\begin{aligned} \frac{\partial l}{\partial \theta_1} &= \frac{1}{\theta_2^2}\sum_{i=1}^{n}(\hat{x}_i - \theta_1) = 0 \\ \frac{\partial l}{\partial \theta_2} &= -\frac{n}{\theta_2} + \frac{1}{\theta_2^3}\sum_{i=1}^{n}(\hat{x}_i - \theta_1 t)^2 = 0 \end{aligned} \tag{5.18}$$

yielding:

$$\begin{aligned} \theta_1 &= \frac{1}{n}\sum_{i=1}^{n}\hat{x}_i \\ \theta_2 &= \sqrt{\frac{\sum_{i=1}^{n}(\hat{x}_i - \theta_1)^2}{n}} \end{aligned} \tag{5.19}$$

which, by using the sample data gives:

$$\theta_1 = \mu = 32.665 \qquad \theta_2 = \sigma = 4.04$$

Not surprisingly, the same result as that obtained using the method of moments is obtained here.

As mentioned previously, the covariance matrix $\mathbf{C}_{\Theta\Theta}$ for the parameters estimates may be determined through the *information matrix* **H** containing the second-order partial derivatives of the log-likelihood function, see Eq. 5.17. The information matrix may be found to be:

$$\mathbf{H} = \begin{pmatrix} \frac{n}{\theta_2^2} & \frac{2\sum_{i=1}^{n}(x_i-\theta_1)}{\theta_2^3} \\ \frac{2\sum_{i=1}^{n}(x_i-\theta_1)}{\theta_2^3} & -\frac{n}{\theta_2^2}+\frac{3\sum_{i=1}^{n}(x_i-\theta_1)^2}{\theta_2^4} \end{pmatrix} \tag{5.20}$$

whereby the covariance matrix is evaluated using the sample data as:

$$\mathbf{C}_{\Theta\Theta} = \mathbf{H}^{-1} = \begin{pmatrix} 0.818 & 0 \\ 0 & 0.409 \end{pmatrix} = \begin{pmatrix} Var[\theta_1] & 0 \\ 0 & Var[\theta_2] \end{pmatrix} \tag{5.21}$$

In probabilistic modeling, where the concrete cube compressive strength enters as a random variable, it is then possible to take into account the statistical uncertainty associated with the estimates of the distribution parameters for the distribution function, simply by including the distribution parameters in the reliability analysis as Normal distributed variables with the evaluated mean values and covariances.

Lecture 9 (Aim of the Present Lecture) This lecture introduces Bayesian estimation methods. The use of these methods makes it possible to update an existing probabilistic model with newly available data. Bayesian regression analysis makes use of this principle. The basic principle here is that the relation between two random variables is represented through a regression line which can be updated based on new data.
On the basis of the lecture it is expected that the reader will acquire knowledge and skills with regard to:

- Which kind of information is typically contained in an a priori model of the probabilistic distribution of a random variable?
- How is the information from new data and the a priori information weighted in Bayesian estimation methods?
- What is a conjugated prior distribution?
- What are the basic assumptions made in linear regression?
- What principle is behind the least squares method?
- How are the uncertainties quantified in a regression model?
- How can a regression model be updated in line with Bayesian regression analysis?

5.4 Bayesian Estimation Methods

A typical situation in risk and reliability analysis is that a *prior* probabilistic model for a random variable is available, e.g. a distribution function and its corresponding distribution parameters have been chosen and estimated based on previous experimental results, experience and professional judgment. As soon as additional data becomes available, it is desired to update the distribution parameters of the *prior* probabilistic model on the basis of this data, weighing the confidence in the prior model consistently with the evidence provided by the new data.

Consider a random variable X with density function $f_X(x)$. If $\boldsymbol{\theta}$ denotes a vector of parameters defining its distribution, the density function of the random variable X can be written as:

$$f_X(x, \boldsymbol{\theta}) \tag{5.22}$$

If X is Normal distributed, then $\boldsymbol{\theta}$ would contain the mean and the standard deviation of X.

If the parameters are uncertain, $f_X(x, \boldsymbol{\theta})$ can be considered as a conditional density function: $f_X(x|\boldsymbol{\theta})$. $\boldsymbol{\theta}$ denotes a realization of the random variable $\boldsymbol{\Theta}$. The initial probability density function for the parameters $\boldsymbol{\Theta}$ is $f'_{\boldsymbol{\Theta}}(\boldsymbol{\theta})$ and is denoted the *prior density function*.

It is assumed that n observations (realizations) of the random variable X are available, thereby building up the sample $\hat{\mathbf{x}} = \left(\hat{x}_1, \hat{x}_2, \ldots, \hat{x}_n\right)^T$. The realizations are assumed to be independent. The updated density function $f''_{\boldsymbol{\Theta}}(\boldsymbol{\theta}|\hat{\mathbf{x}})$ of the uncertain parameters $\boldsymbol{\Theta}$, given the realizations $\hat{\mathbf{x}}$, is denoted the *posterior density function* and is given by (see JCSS [8]):

$$f''_{\boldsymbol{\Theta}}(\boldsymbol{\theta}|\hat{\mathbf{x}}) = \frac{L(\hat{\mathbf{x}}|\boldsymbol{\theta})(\hat{\mathbf{x}}|\boldsymbol{\theta}) f'_{\boldsymbol{\Theta}}(\boldsymbol{\theta})}{\int L(\hat{\mathbf{x}}|\boldsymbol{\theta}) f'_{\boldsymbol{\Theta}}(\boldsymbol{\theta})\, d\boldsymbol{\theta}} \tag{5.23}$$

where $L(\hat{\mathbf{x}}|\boldsymbol{\theta}) = \prod_{i=1}^{n} f_X(\hat{x}_i\,|\,\boldsymbol{\theta})$ is the likelihood corresponding to the given observations assuming that the distribution parameters are $\boldsymbol{\theta}$. The integration in Eq. 5.23 is over all possible realizations of $\boldsymbol{\Theta}$. The updated density function of the random variable X given the realization $\hat{\mathbf{x}}$ is denoted the *predictive density function* and is defined by:

$$f_X(x|\hat{\mathbf{x}}) = \int f_X(x|\boldsymbol{\theta}) f''_{\boldsymbol{\Theta}}(\boldsymbol{\theta}|\hat{\mathbf{x}})\, d\boldsymbol{\theta} \tag{5.24}$$

Given the distribution function for the random variable X, the prior distribution is often chosen such that the posterior distribution will be of the same type as the prior distribution (*conjugated prior*). In the literature, a number of prior, posterior and predictive distribution functions can be found, see e.g. JCSS [8]. Analytical solutions concerned with the following problems can be found for:

- Normal distribution with unknown mean
- Normal distribution with unknown standard deviation
- Normal distribution with unknown mean and standard deviation

- Gumbel distribution
- Weibull distribution
- Exponential distribution
- Bernoulli distribution
- Poisson distribution
- Multidimensional Normal distribution with unknown means
- Multidimensional Normal distribution with unknown standard deviations
- Multidimensional Normal distribution with unknown means and standard deviations

The parameters in the *prior distribution* can be chosen or calculated in such a way that the prior reflects:

- the known (initial) observations of the random variable X from which estimates of the parameters in the prior distribution can be calculated, and/or
- the subjective knowledge on the distribution of the parameters $\boldsymbol{\Theta}$ in the distribution of X.

In this way, it is possible to choose a prior distribution reflecting a range of situations from a very good prior knowledge of the parameters (small standard deviation) to almost no knowledge of the parameters (large standard deviation).

5.4.1 Example 5.2—Yield Stress of a Steel Bar

As an example on the updating of random variables consider the probabilistic modeling of the yield stress of steel. The *prior* probabilistic model for the yield stress is assumed to be Normal distributed with known (deterministic) standard deviation $\sigma_{f_y} = 17.5$ MPa and uncertain mean value μ_{f_y}. The mean value μ_{f_y} is assumed to be Normal distributed with a mean value $\mu' = 350$ MPa and standard deviation $\sigma' = 10$ MPa.

Assume now that yield stress tests are performed on 5 steel samples taken from a batch of the same steel material. The results of the yield stress tests are $\hat{\mathbf{f}}_{\mathbf{y}} = (365, 347, 354, 362, 348)^T$.

Based on the test results, the *prior* probabilistic model for the mean value of the yield stress can be updated using natural conjugate probability distributions as mentioned earlier.

In the case considered with a Normal distributed random variable with uncertain mean and known standard deviation, the posterior probability density as given in Eq. 5.23 may be reduced to (JCSS [8])

$$f''(\mu_{f_y}) = \frac{1}{\sqrt{2\pi}\sigma''} \exp\left(-\frac{1}{2}\left(\frac{\mu_{f_y} - \mu''}{\sigma''}\right)^2\right) \tag{5.25}$$

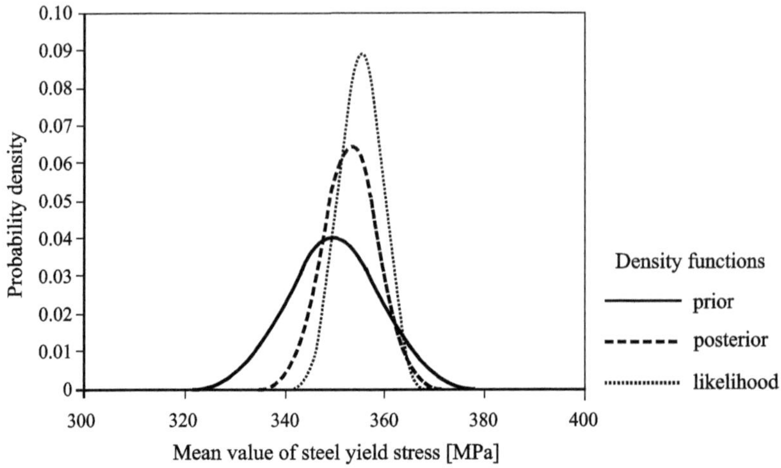

Fig. 5.6 Illustration of *prior* and *posterior* probability density and likelihood functions for the mean value of the steel yield stress

where

$$\mu'' = \frac{\frac{\mu'}{n} + \frac{\bar{x}}{n'}}{\frac{1}{n'} + \frac{1}{n}} \tag{5.26}$$

and

$$\sigma'' = \sqrt{\frac{\frac{\sigma_{f_y}^2}{n'} \cdot \frac{\sigma'^2}{n}}{\frac{\sigma'^2}{n'} + \frac{\sigma'^2}{n}}} \tag{5.27}$$

and

$$n' = \frac{\sigma_{f_y}^2}{\sigma'^2} \tag{5.28}$$

$\bar{x}$ is the sample mean of the observations, n' is the sample size assumed for the *prior* distribution of μ_{f_y} and n is the sample size for the new sample. In the present example $n' = 3.06$. Based on the new observations, the *posterior* parameters are $\mu'' = 353.22$ and $\sigma'' = 6.16$.

Figure 5.6 illustrates the *prior* and the *posterior* probability density functions for μ_{f_y}.

The likelihood of the observation can be established as:

$$L(\mu_{f_y} | \hat{\mathbf{f}}_{\mathbf{y}}) = \prod_{1}^{5} \frac{1}{\sqrt{2\pi}\sigma_{f_y}} \exp\left(-\frac{1}{2}\frac{(\hat{f}_{yi} - \mu_{f_y})^2}{\sigma_{f_y}{}^2}\right) \tag{5.29}$$

The likelihood function is also shown in Fig. 5.6. It is seen from Fig. 5.6 that the effect of the test results can be quite significant. According to JCSS [8] the *predictive*

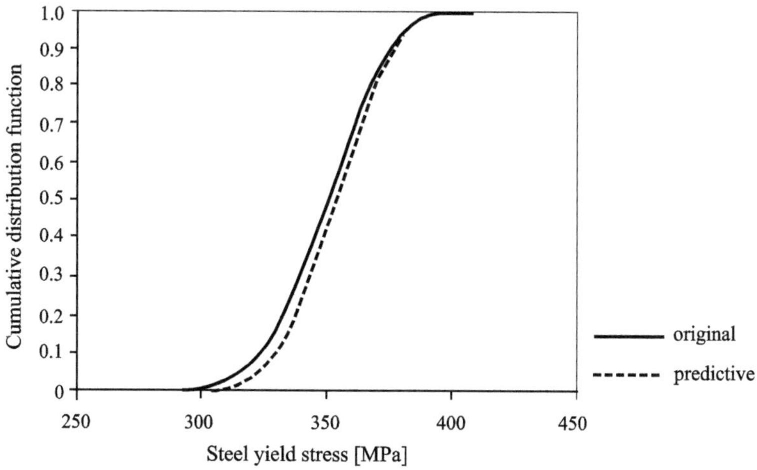

Fig. 5.7 Illustration of original and predictive probability distribution function for the steel yield stress

probability density function for the steel yield stress may be determined as:

$$f_{f_y}(f_y|\hat{\mathbf{f}}_{\mathbf{y}}) = \frac{1}{\sqrt{2\pi}\sigma'''}\exp\left(-\frac{1}{2}\left(\frac{f_y-\mu''}{\sigma'''}\right)^2\right) \tag{5.30}$$

where

$$\sigma'''^2 = \sigma''^2 + \sigma_{f_y}^2 \tag{5.31}$$

In Fig. 5.7 the predictive probability distribution and the probability distribution function for the steel yield stress based on the *prior* information of the mean value are shown.

The 5% quantile, which is a typical characteristic value for the steel yield stress, has changed from 317 MPa to 322 MPa as a result of the Bayesian updating process (see Fig. 5.7).

5.5 Bayesian Regression Analysis

Bayesian regression analysis is a very useful modeling tool in engineering where (semi-)empirical relationships based on experimental evidence play a big role. In this section, only the special, but in practice useful, case of *univariate linear models* is considered. A probabilistic model for a random variable Y is to be developed where Y is a linear function of a random variable X. The relationship between X and Y can be expressed as follows:

$$Y = \beta_0 + \beta_1 X + \varepsilon \tag{5.32}$$

where β_0 and β_1 are model parameters, $i \in \{1, 2, \ldots, n\}$ is the index of test result, and ε_i is a realization of the residual ε. It is assumed that the residual ε is Normal distributed with zero mean and standard deviation σ_ε. Hence it is seen that Y for given x is also Normal distributed. It is assumed that n experiment or test results are available $(\hat{\mathbf{x}}, \hat{\mathbf{y}})^T$. The first step involves assessing the *prior* distribution of the model parameters. In the next step, the model is updated with new measurements.

5.5.1 Linear Regression: Prior Model

The model parameters β_0 and β_1 can be assessed by using the *least squares method*. For this, the model error ε_i is assessed for each point by determining the difference between the measured and predicted values of y_i. The model parameters which minimize the sum of all squared error terms are then obtained. The squared error terms are used in order to account for positive and negative errors equally.

$$\min \sum_{i=1}^{n} \varepsilon_i^2 = \min \sum_{i=1}^{n} (\hat{y}_i - (\beta_0 + \beta_1 \hat{x}_i))^2 \tag{5.33}$$

To minimize this function we set its partial derivatives with respect to β_0 and β_1 to be equal to zero:

$$\begin{aligned} 0 &= \sum_{i=1}^{n} (\hat{y}_i - \beta_0 - \beta_1 \hat{x}_i)^2 \frac{\partial}{\partial \beta_0} = -2 \sum_{i=1}^{n} (y_i - \beta_0 - \beta_1 x_i) \\ 0 &= \sum_{i=1}^{n} (\hat{y}_i - \beta_0 - \beta_1 \hat{x}_i)^2 \frac{\partial}{\partial \beta_1} = -2 \sum_{i=1}^{n} (\hat{y}_i - \beta_0 - \beta_1 \hat{x}_i)\hat{x}_i \end{aligned} \tag{5.34}$$

The set of equations obtained from Eq. 5.34 can be written as:

$$\sum_{i=1}^{n} \hat{y}_i = \beta_1 \sum_{i=1}^{n} \hat{x}_i + \beta_0 \sum_{i=1}^{n} 1 \tag{5.35}$$

$$\sum_{i=1}^{n} \hat{x}_i \hat{y}_i = \beta_1 \sum_{i=1}^{n} \hat{x}_i^2 + \beta_0 \sum_{i=1}^{n} \hat{x}_i \tag{5.36}$$

This equation system can be described in matrix notation:

$$\begin{pmatrix} \mathbf{1}^T \times \hat{\mathbf{y}}^T \\ \hat{\mathbf{x}}^T \times \hat{\mathbf{y}}^T \end{pmatrix} = \begin{pmatrix} \hat{\mathbf{x}}^T \times \mathbf{1}^T & \mathbf{1}^T \times \mathbf{1}^T \\ \hat{\mathbf{x}}^T \times \hat{\mathbf{x}}^T & \hat{\mathbf{x}}^T \times \mathbf{1}^T \end{pmatrix} \begin{pmatrix} \beta_1 \\ \beta_0 \end{pmatrix} \tag{5.37}$$

or condensed:

$$\hat{\mathbf{X}}^T \hat{\mathbf{y}} = \hat{\mathbf{X}}^T \hat{\mathbf{X}} \boldsymbol{\beta} \tag{5.38}$$

The vector $\hat{\mathbf{y}}$ contains the test results $\hat{y}_i$ while the matrix $\mathbf{X}$ contains ones in the first column and the test results $\hat{x}_i$ in the second column. Equation 5.38 can be reformulated as:

$$\boldsymbol{\beta} = \left(\hat{\mathbf{X}}^T\hat{\mathbf{X}}\right)^{-1}\hat{\mathbf{X}}^T\hat{\mathbf{y}} \tag{5.39}$$

where:

$$\hat{\mathbf{X}}^T\hat{\mathbf{X}} = \begin{pmatrix} 1 & \cdots & 1 \\ \hat{x}_1 & \cdots & \hat{x}_n \end{pmatrix} \begin{pmatrix} 1 & \hat{x}_1 \\ \vdots & \vdots \\ 1 & \hat{x}_n \end{pmatrix} = \begin{pmatrix} n \cdot 1^2 & \sum_{i=1}^n \hat{x}_i \\ \sum_{i=1}^n \hat{x}_i & \sum_{i=1}^n \hat{x}_i^2 \end{pmatrix} \tag{5.40}$$

and

$$\hat{\mathbf{X}}^T\hat{\mathbf{y}} = \begin{pmatrix} 1 & \cdots & 1 \\ \hat{x}_1 & \cdots & \hat{x}_n \end{pmatrix} \begin{pmatrix} \hat{y}_1 \\ \vdots \\ \hat{y}_n \end{pmatrix} = \begin{pmatrix} \sum_{i=1}^n \hat{y}_i \\ \sum_{i=1}^n \hat{x}_i \hat{y}_i \end{pmatrix} \tag{5.41}$$

Equation 5.39 can also be derived using the Method of Maximum Likelihood. In a simple linear regression, the following solutions for β_0 and β_1 exist:

$$\beta_0 = \frac{1}{n}\sum_{i=1}^n \hat{y}_i - \beta_1 \frac{1}{n}\sum_{i=1}^n \hat{x}_i = \bar{y} - \beta_1 \bar{x} \tag{5.42}$$

$$\beta_1 = \frac{\frac{1}{n}\sum_{i=1}^n \hat{y}_i\hat{x}_i - \bar{y}\frac{1}{n}\sum_{i=1}^n \hat{x}_i}{\frac{1}{n}\sum_{i=1}^n \hat{x}_i^2 - \bar{x}\frac{1}{n}\sum_{i=1}^n \hat{x}_i} = \frac{\frac{1}{n}\sum_{i=1}^n \hat{y}_i\hat{x}_i - \bar{y}\bar{x}}{\frac{1}{n}\sum_{i=1}^n \hat{x}_i^2 - \bar{x}^2} = \frac{s_{XY}}{s_X^2} \tag{5.43}$$

The standard deviation of the error term ε is assessed by:

$$\sigma_\varepsilon = \sqrt{\frac{\sum_{i=1}^n \varepsilon_i^2}{n-k}} \tag{5.44}$$

where n is the number of measurements and k is the number of parameters in $\boldsymbol{\beta}$. The variance of the error term can be written in matrix form as:

$$\sigma_\varepsilon^2 = \left(\mathbf{y} - \hat{\mathbf{X}}\boldsymbol{\beta}\right)^T\left(\mathbf{y} - \hat{\mathbf{X}}\boldsymbol{\beta}\right)/(n-k) \tag{5.45}$$

The conditional distribution of $\hat{Y}$ given $\boldsymbol{\beta}$ and $\hat{\mathbf{X}}$ follows a Normal distribution with the following parameters:

$$E\left(\hat{Y}|\boldsymbol{\beta};\hat{\mathbf{X}}\right) = \boldsymbol{\beta}\hat{\mathbf{X}} \tag{5.46}$$

$$Var\left(\hat{Y}|\boldsymbol{\beta};\hat{\mathbf{X}}\right) = \sigma_\varepsilon^2 \tag{5.47}$$

The uncertainty of the parameters $\boldsymbol{\beta}$ can be represented using the covariance matrix, where the parameter variances are contained in the diagonal:

$$Cov(\boldsymbol{\beta}) = \sigma_\varepsilon^2 \mathbf{V}_\beta \quad \text{with } \mathbf{V}_\beta = \left(\hat{\mathbf{X}}^T\hat{\mathbf{X}}\right)^{-1} \tag{5.48}$$

In case the random variable $\hat{Y}$ needs to be modeled using not only one, but k components in $\hat{\mathbf{X}}$, the model can be generalized to:

$$y_i = \beta_o + \sum_{j=1}^{k} x_{ij}\beta_j + \varepsilon_i \tag{5.49}$$

The parameters $\boldsymbol{\beta}$ can also be assessed through Eq. 5.39 with the only difference being that the matrix $\mathbf{X}$ holds one column for each of r components.

5.5.2 Example 5.3—Tensile Strength of Timber: Prior Model

In the grading of timber materials it is normal to classify the sawn timber tensile strength based on the tensile modulus of elasticity. To be able to do this it is necessary to determine the correlation between the tensile strength X and the tensile modulus of elasticity Y on the basis of measured values.

Using the experiment results from Table 5.2 the parameters β_0 and β_1 can now be assessed for the model $Y = \beta_0 + \beta_1 X + \varepsilon$:

$$\hat{\mathbf{X}} = \begin{pmatrix} 1 & 19.11 \\ 1 & 12.30 \\ 1 & 14.83 \\ 1 & 13.18 \end{pmatrix} \qquad \hat{\mathbf{y}} = \begin{pmatrix} 8426 \\ 7092 \\ 7347 \\ 7917 \end{pmatrix}$$

$$\begin{aligned}
\boldsymbol{\beta} &= \left(\hat{\mathbf{X}}^T\hat{\mathbf{X}}\right)^{-1}\hat{\mathbf{X}}^T\hat{\mathbf{y}} \\
&= \left[\begin{pmatrix} 1 & 1 & 1 & 1 \\ 19.11 & 12.30 & 14.83 & 13.18 \end{pmatrix}\begin{pmatrix} 1 & 19.11 \\ 1 & 12.30 \\ 1 & 14.83 \\ 1 & 13.18 \end{pmatrix}\right]^{-1} \\
&\quad \times \begin{pmatrix} 1 & 1 & 1 & 1 \\ 19.11 & 12.30 & 14.83 & 13.18 \end{pmatrix}\begin{pmatrix} 8426 \\ 7092 \\ 7347 \\ 7917 \end{pmatrix} \\
&= \begin{pmatrix} 366.19 & 236.05 & 284.40 & 252.87 \\ 236.05 & 152.29 & 183.41 & 163.11 \\ 284.40 & 183.41 & 220.93 & 196.46 \\ 252.87 & 163.11 & 196.46 & 174.71 \end{pmatrix}^{-1}\begin{pmatrix} 30782 \\ 461554 \end{pmatrix} \\
&= \begin{pmatrix} 8.29 & -0.54 \\ -0.54 & 0.04 \end{pmatrix}\begin{pmatrix} 30782 \\ 461554 \end{pmatrix} = \begin{pmatrix} 5374.12 \\ 156.27 \end{pmatrix} = \begin{pmatrix} \beta_0 \\ \beta_1 \end{pmatrix}
\end{aligned}$$

Table 5.2 Measured values of tensile strength and tensile modulus of elasticity of sawn timber

Tensile strength $\hat{\mathbf{x}}$ [MPa]	Tensile modulus of elasticity $\hat{\mathbf{y}}$ [MPa]
19.11	8426
12.30	7092
14.83	7347
13.18	7917

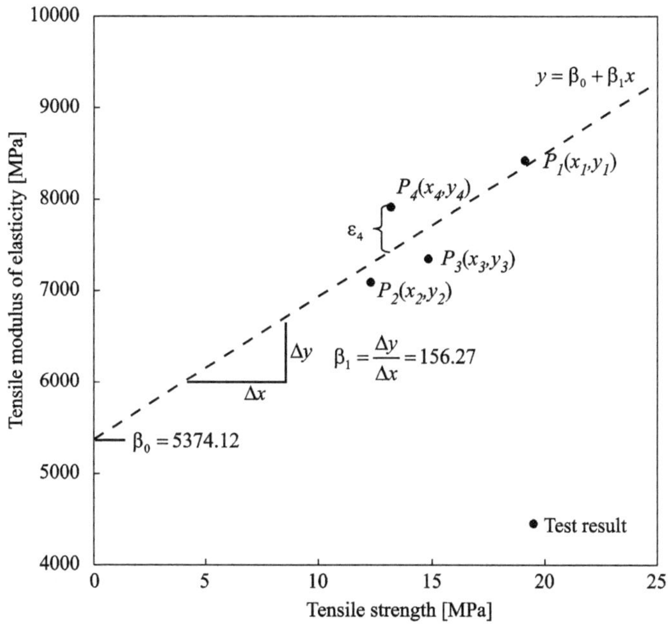

Fig. 5.8 Regression model and test results

The standard deviation of the error term can be determined as follows:

$$\sum_{i=1}^{n} \varepsilon_i^2 = \sum_{i=1}^{n} (\hat{y}_i - (\beta_0 + \beta_1 \hat{x}_i))^2$$

$$= (8426 - (5374.12 + 156.27 \cdot 19.11))^2 + \cdots$$

$$+(7917 - (5374.12 + 156.27 \cdot 13.18))^2$$

$$= 39'8287$$

$$\sigma_\varepsilon = \sqrt{\frac{\sum_{i=1}^{n} \varepsilon_i^2}{n-k}} = \sqrt{\frac{\sum_{i=1}^{n} \varepsilon_i^2}{n-2}} = \sqrt{\frac{39'8287}{4-2}} = 446.25 \text{ MPa}$$

Hence, the regression equation is (see Fig. 5.8):

$$Y = 5374.12 + 156.27X + \varepsilon \quad \varepsilon \sim N(0, 446.25)$$

The uncertainties of the parameters $\boldsymbol{\beta}$ are determined in the following way:

$$\mathbf{V}_\beta = \left(\hat{\mathbf{X}}^T\hat{\mathbf{X}}\right)^{-1}$$

$$= \left[\begin{pmatrix} 1 & 1 & 1 & 1 \\ 19.11 & 12.30 & 14.83 & 13.18 \end{pmatrix}\begin{pmatrix} 1 & 19.11 \\ 1 & 12.30 \\ 1 & 14.83 \\ 1 & 13.18 \end{pmatrix}\right]^{-1} \begin{pmatrix} 8.29 & -0.54 \\ -0.54 & 0.04 \end{pmatrix}$$

$$Cov\,(\boldsymbol{\beta}) = \sigma_\varepsilon^2 \mathbf{V}_\beta = 446.25 \begin{pmatrix} 8.29 & -0.54 \\ -0.54 & 0.04 \end{pmatrix} = \begin{pmatrix} 3'700.4 & -241.6 \\ -241.6 & 16.3 \end{pmatrix}$$

5.5.3 Updating Regression Coefficients: Posterior Model

In the foregoing section, linear regression has been introduced as a useful tool to investigate the linear relationship between two random variables. This section shows how the *prior* model for the regression coefficients $\boldsymbol{\beta}$ can be updated based on new measurements. It is assumed that in addition to the parameters $\boldsymbol{\beta}'$ and $\mathbf{V}'_\beta$ from the prior model, the new measurements $\mathbf{y}$ and $\mathbf{X}$ are given.

The parameters for the *posterior* model are then derived from the following equation:

$$(\mathbf{V}''_\beta)^{-1} = (\mathbf{V}'_\beta)^{-1} + \hat{\mathbf{X}}^T\hat{\mathbf{X}} \tag{5.50}$$

$$\boldsymbol{\beta}'' = \mathbf{V}''_\beta\left((\mathbf{V}'_\beta)^{-1}\boldsymbol{\beta}' + \hat{\mathbf{X}}^T\hat{\mathbf{y}}\right) \tag{5.51}$$

5.5.4 Example 5.4—Updating Regression Coefficients (Determined in Example 5.3)

The *prior* model in Example 5.3 describes the correlation between the tensile strength X and the tensile modulus of elasticity Y of timber:

$$\boldsymbol{\beta}' = \begin{pmatrix} 5374.12 \\ 156.27 \end{pmatrix} \qquad \mathbf{V}'_\beta = \begin{pmatrix} 8.29 & -0.54 \\ -0.54 & 0.04 \end{pmatrix}$$

Now, the *prior* model is updated with two new test results, as indicated in Table 5.3.

$$\mathbf{X} = \begin{pmatrix} 1 & 18.04 \\ 1 & 24.10 \end{pmatrix} \qquad \mathbf{y} = \begin{pmatrix} 7581 \\ 9661 \end{pmatrix}$$

$$(\mathbf{V}''_\beta)^{-1} = (\mathbf{V}'_\beta)^{-1} + \hat{\mathbf{X}}^T\hat{\mathbf{X}}$$

$$= \begin{pmatrix} 8.29 & -0.54 \\ -0.54 & 0.04 \end{pmatrix} + \mathbf{X}^T\mathbf{X}$$

$$= \begin{pmatrix} 8.29 & -0.54 \\ -0.54 & 0.04 \end{pmatrix} + \left[\begin{pmatrix} 1 & 1 \\ 18.04 & 24.10 \end{pmatrix}\begin{pmatrix} 1 & 18.04 \\ 1 & 24.10 \end{pmatrix}\right]$$

Table 5.3 Additional test results used for updating

Tensile strength $\hat{\mathbf{x}}$ [MPa]	Tensile modulus of elasticity $\hat{\mathbf{y}}$ [MPa]
18.04	7581
24.10	9661

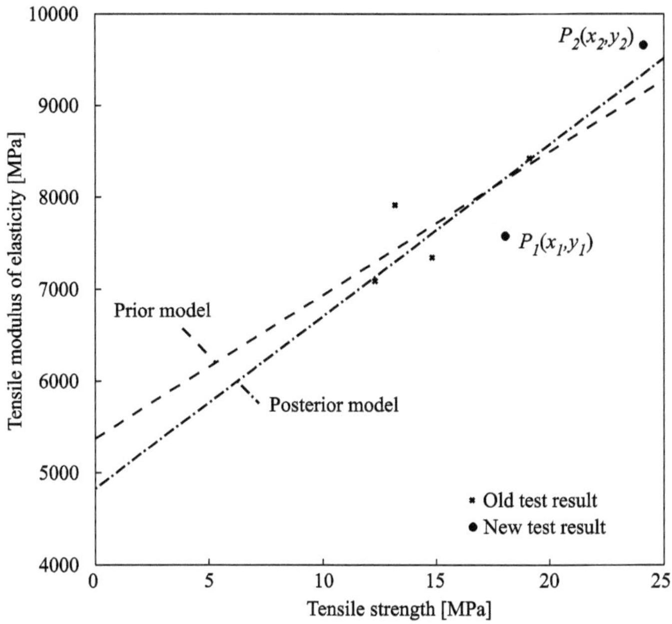

Fig. 5.9 Updated regression model

$$= \begin{pmatrix} 8.29 & -0.54 \\ -0.54 & 0.04 \end{pmatrix} + \begin{pmatrix} 2 & 42.14 \\ 42.14 & 906.25 \end{pmatrix} = \begin{pmatrix} 10.29 & 41.6 \\ 41.6 & 906.29 \end{pmatrix}$$

$$\boldsymbol{\beta}'' = \mathbf{V}''_{\beta}\left((\mathbf{V}'_{\beta})^{-1}\boldsymbol{\beta}' + \hat{\mathbf{X}}^T\hat{\mathbf{y}}\right)$$

$$= \begin{pmatrix} 10.29 & 41.6 \\ 41.6 & 906.29 \end{pmatrix}^{-1} \left[\begin{pmatrix} 8.29 & -0.54 \\ -0.54 & 0.04 \end{pmatrix}^T \begin{pmatrix} 5374.12 \\ 156.27 \end{pmatrix}^T \right.$$

$$\left. + \begin{pmatrix} 1 & 1 \\ 18.04 & 24.10 \end{pmatrix} \begin{pmatrix} 7581 \\ 9661 \end{pmatrix} \right]$$

$$= \begin{pmatrix} 3.11 & -0.17 \\ -0.17 & 0.01 \end{pmatrix} \left[\begin{pmatrix} 30782 \\ 461554 \end{pmatrix} + \begin{pmatrix} 17242 \\ 369591 \end{pmatrix} \right]$$

$$= \begin{pmatrix} 3.11 & -0.17 \\ -0.17 & 0.01 \end{pmatrix} \begin{pmatrix} 48024 \\ 831159 \end{pmatrix} = \begin{pmatrix} 4827.58 \\ 187.66 \end{pmatrix} = \begin{pmatrix} \beta''_0 \\ \beta''_1 \end{pmatrix}$$

Hence, the updated regression equation is (see Fig. 5.9):

$$Y = 4827.58 + 187.66X + \varepsilon$$

Lecture 10 (Aim of the Present Lecture) The aim of the present lecture is to introduce the concept of testing for statistical significance. First, it is explained how the statistical properties of sample characteristics depend on the sample size, i.e. the amount of data available, and in this way, the quantification of statistical uncertainty is addressed. Then it is shown how hypotheses related to the statistical properties of probabilistic models may be tested on the basis of data.

On the basis of the lecture it is expected that the reader will acquire knowledge and skills with regard to:

- What are the distribution functions applied in sample statistics and what are the principles on the basis of which they are derived?
- What does it mean that the Chi-square distribution is regenerative?
- How is the Chi-distribution related to the law of Pythagoras?
- How does the expected value of the sample mean depend on the number of samples?
- How does the variance of the sample mean depend on the number of samples?
- How can an unbiased estimator for the sample variance be established?
- What is a confidence interval and how can it be established?
- What is a hypothesis and how can it be tested?
- What is a null-hypothesis and what is an alternate hypothesis?
- What is a type I error and what is a type II error?
- What is the meaning of statistical significance?
- How can tests of the mean of a random variable be performed?

5.6 Probability Distributions in Statistics

Throughout the classical statistical theory some distribution functions are repeatedly used for assessment and testing purposes. These include the important *Chi-square distribution*, the *Chi-distribution*, the *t-distribution* and the *F-distribution*. Here, only the first two distributions are briefly introduced in accordance with Benjamin and Cornell [4]. The distributions are all related and may be derived from the Normal distribution as shown in e.g. Benjamin and Cornell [4]. The numerical evaluation of the distributions may be performed using standard commercial spread sheets such as e.g. Microsoft Excel or tabulations as given in Appendix C.

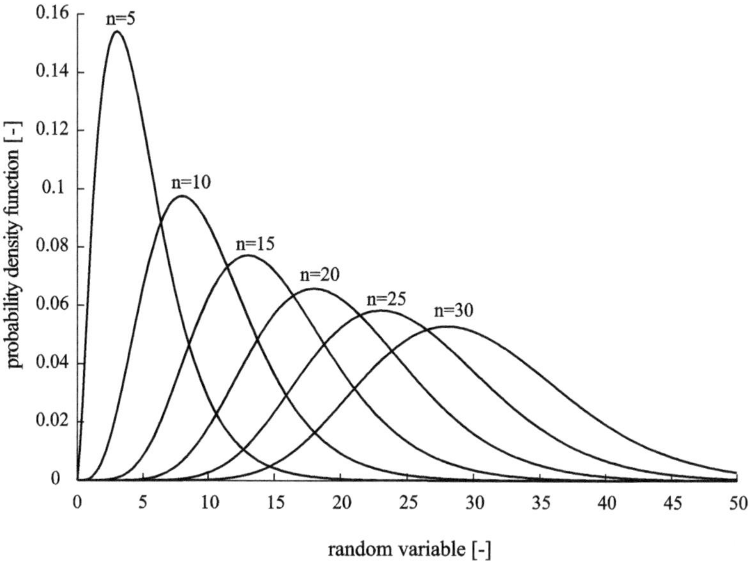

Fig. 5.10 Chi-square distributions with different degrees of freedom (n). With increasing n the chi-square distribution approximates a normal distribution

5.6.1 The Chi-Square (χ^2)-Distribution

When $X_i, i = 1, 2, \ldots, n$ are standard Normal distributed independent random variables the sum of the squares of the random variables Y_n i.e.:

$$Y_n = \sum_{i=1}^{n} X_i^2 \tag{5.52}$$

is said to be Chi-square distributed (some times written as χ^2-distributed) with probability density function:

$$f_{Y_n}(y_n) = \frac{y_n^{(n/2-1)}}{2^{n/2}\Gamma(n/2)} \exp(-y_n/2), \quad y_n \geq 0 \tag{5.53}$$

with mean value $\mu_{Y_n} = n$ (also referred to as the *degrees of freedom*) and variance $\sigma_{Y_n}^2 = 2n$. In Eq. 5.53 $\Gamma(\cdot)$ is the complete *Gamma function* defined by:

$$\Gamma(x) = \int_0^{\infty} e^{-t} t^{x-1}\, dt \tag{5.54}$$

As it shall be seen later, the Chi-square distribution is often applied for assessing the statistical characteristics of squared errors but can also be applied in various engineering assessments involving squares of Normal distributed variables such as e.g. the drag component of wave and wind loads and kinetic energy components.

The Chi-square distribution is regenerative in the sense that the sum of two Chi-square distributed variables i.e. $Y_{n_1} + Y_{n_2}$ is also Chi-square distributed with $n_1 + n_2$ degrees of freedom.

From the additive character of the Chi-square distribution (Eq. 5.52), it is seen, from application of the central limit theorem, that for sufficiently large n the Chi-square distribution converges towards a Normal distribution with mean value $\mu_{Y_n} = n$ and variance $\sigma_{Y_n}^2 = 2n$, see also Fig. 5.10.

5.6.2 *The Chi (χ)-Distribution*

When a random variable Z is given as the square root of a Chi-square distributed random variable Y_n, the variable Z is said to follow a *Chi-distribution* (sometimes written as χ-distributed) with probability density function:

$$f_Z(z) = \frac{z^{(n-1)}}{2^{n/2-1}\Gamma(n/2)} \exp(-z^2/2), \quad z \geq 0 \tag{5.55}$$

The mean value μ_z and the variance σ_z^2 are given by:

$$\mu_z = \sqrt{2}\frac{\Gamma((n+1)/2)}{\Gamma(n/2)} \tag{5.56}$$

$$\sigma_z^2 = n - 2\frac{\Gamma^2((n+1)/2)}{\Gamma^2(n/2)} \tag{5.57}$$

The Chi-distribution is e.g. used for the assessment of the distances measured using the principles of Pythagoras or Euclidean norms, and for the assessment of the statistical characteristics of standard deviations.

5.7 Estimators for Sample Descriptors—Sample Statistics

When frequentistic information becomes available e.g. in the form of experimental results, a first step is often to try to assess the data simply as they are, without too many assumptions regarding the probabilistic characteristics of the mechanism/process which generated them. Such an assessment typically concerns the numerical summaries as described in Chap. 3, e.g. the sample moments, but could in principle be any *sample characteristic* of the observed data which is found to be of interest in a given situation. In statistical terms, such characteristics are called *sample statistics*, and in the following section the statistical characteristics of such sample statistics will be considered in some detail. To this end, the uncertainty associated with parameter estimators will be assessed and *confidence intervals* on the estimators will be introduced. Finally, *significance testing* is introduced as a means of evaluating the significance of the variability of statistical data.

5.7.1 Statistical Characteristics of the Sample Average

As an example, consider the case where the permeability of a particular soil is of interest in an engineering decision problem. Due to various effects such as inherent natural variability in the soil composition and the consolidation, the permeability of the considered soil is associated with significant uncertainty. As an engineering model, it is assumed that this uncertainty can be taken into account in the formulation of the decision problem by modeling the permeability by a random variable X with distribution function $F_X(x;\mathbf{p})$. Having selected the family of distribution functions, i.e. the distribution function $F_X(x)$, it is still needed to estimate the parameters $\mathbf{p}$, and as will be seen in the following sections, this can be done by the method of moments or the maximum likelihood method, provided that experiment results $\hat{\mathbf{x}} = (\hat{x}_1, \hat{x}_2, \ldots, \hat{x}_n)^T$ are available.

In order to better appreciate the uncertainty associated with statistical characteristics such as distribution parameters $\mathbf{p}$, the statistical properties of these parameters are now considered. It is assumed that experiment results of the yet unknown values are collected in the vector $\mathbf{X}$. If the experiments are conducted independently, the realizations can be modeled as independent random variables $X_i, i = 1, 2, \ldots, n$ with cumulative distribution functions $F_{X_i}(x_i;\mathbf{p}) = F_X(x;\mathbf{p})$, $i = 1, 2, \ldots, n$. Based on the probabilistic model of the realizations X_i, $i = 1, 2, \ldots, n$ it is possible to assess the statistical characteristics of the unknown *sample mean* $\bar{X}$ and the unknown *sample variance* S^2 given by:

$$\bar{X} = \frac{1}{n}\sum_{i=1}^{n} X_i \tag{5.58}$$

$$S^2 = \frac{1}{n}\sum_{i=1}^{n} (X_i - \bar{X})^2 \tag{5.59}$$

The sample mean $\bar{X}$ and the sample variance S^2 are random variables given in terms of functions of the experiment outcomes X_i, $i = 1, 2, \ldots, n$. Such functions are in general called *sample statistics* and include, as mentioned previously, any characteristic of the considered distribution of interest.

In order to assess the uncertainty by which the sample mean $\bar{X}$ is associated, it is interesting to consider its expected value $E[\bar{X}]$ and variance $Var[\bar{X}]$. The expected value is given as:

$$E[\bar{X}] = E\left[\frac{1}{n}\sum_{i=1}^{n} X_i\right] = \frac{1}{n}\sum_{i=1}^{n} E[X_i] = \frac{1}{n} n\mu_X = \mu_X \tag{5.60}$$

which shows that the expected value of the sample mean is indeed equal to the expected value of the underlying random variable, in this case the soil permeability. It was expected that the sample mean is a good estimator for the expected value of

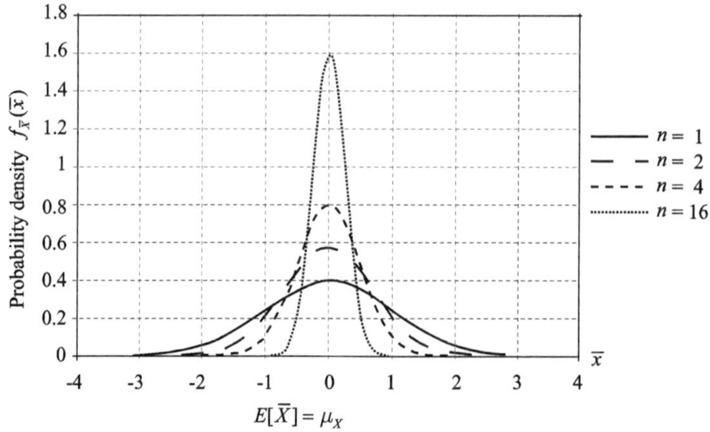

Fig. 5.11 Illustration of the probability density function of a sample average for different sample sizes n

a random variable. However, due to the fact that the sample mean would be a realization of a random variable, it is clear that the sample mean will normally not turn out to be exactly equal to the expected value of the underlying random variable.

The variability of the sample mean around its expected value can be assessed through the variance of the sample mean $Var[\bar{X}]$ given by:

$$Var[\bar{X}] = E\left[(\bar{X} - \mu_X)^2\right] = E\left[(\bar{X} - E[\bar{X}])^2\right] \tag{5.61}$$

Equation 5.61 may be rewritten as:

$$\begin{aligned} Var[\bar{X}] &= Var\left[\frac{1}{n}\sum_{i=1}^{n} X_i\right] = \frac{1}{n^2} Var\left[\sum_{i=1}^{n} X_i\right] = \frac{1}{n^2}\sum_{i=1}^{n} Var[X_i] \\ &= \frac{1}{n^2}\sum_{i=1}^{n} E\left[(X_i - \mu_X)^2\right] = \frac{1}{n}\sigma_X^2 \end{aligned} \tag{5.62}$$

from which it is seen that the variance of the sample mean $Var[\bar{X}]$ decreases linearly as a function of the number of samples. Considering the probability that the sample mean $\bar{X}$ will lie within a certain range around the expected value of X i.e. $\mu_X \pm k\sigma_X$, it is seen from Eq. 5.62 that the *band width factor* k may be reduced by a factor of 2 by increasing the number of experiments by a factor of 4. To reduce k by a factor of 4, the number of experiments must be increased by a factor of 16. It is seen that it becomes increasingly expensive in terms of experiments to reduce the uncertainty associated with the sample mean. In Fig. 5.11 the probability density function of a sample mean is illustrated for different values of the sample size n.

5.7.2 Statistical Characteristics of the Sample Variance

Whereas the *sample average* is of interest as an estimator of the mean value μ_X of a random variable, the sample variance S^2 is of interest as an estimator of the variance σ_X^2. The expected value of the *sample variance* is determined by taking the expectation of the sample variance as given by Eq. 5.59, i.e.:

$$\begin{aligned}
E[S^2] &= E\left[\frac{1}{n}\sum_{i=1}^{n}(X_i - \bar{X})^2\right] = \frac{1}{n}E\left[\sum_{i=1}^{n}((X_i - \mu_X) - (\bar{X} - \mu_X))^2\right] \\
&= \frac{1}{n}\left(\sum_{i=1}^{n}E\left[(X_i - \mu_X)^2\right] - nE\left[(\bar{X} - \mu_X)^2\right]\right) \\
&= \frac{1}{n}\left(nE\left[(X_i - \mu_X)^2\right] - nE\left[(\bar{X} - \mu_X)^2\right]\right) \\
&= \frac{1}{n}\left(n\sigma_X^2 - n\frac{\sigma_X^2}{n}\right) = \sigma_X^2 - \frac{1}{n}\sigma_X^2 = \frac{(n-1)}{n}\sigma_X^2
\end{aligned} \tag{5.63}$$

In the step going from the third line to the fourth line in Eq. 5.63 the assumption of independence has been used, i.e. using that $E[X_i X_j] = 0$ for $i \neq j$. From Eq. 5.63 it is noticed that the expected value of the sample variance is different from the variance of the underlying random variable. Even though this difference is small for large sample sizes n, this is disturbing and essentially means that the estimator S^2 is *biased*, i.e. its mean value is different from σ_X^2. An estimator of the variance σ_X^2 which is *unbiased* S^2_{unbiased} may, however, easily be constructed from S^2 as:

$$S^2_{\text{unbiased}} = \frac{n}{n-1}S^2 = \frac{1}{n-1}\sum_{i=1}^{n}(X_i - \bar{X})^2 \tag{5.64}$$

It is noted (refer to Sect. 5.3) that the biased estimator S^2 for the variance is applied in both the methods of moments and the maximum likelihood method.

The goodness of an estimator cannot, however, be judged alone on the basis of whether or not it is biased. Another characteristic of estimators commonly used is the efficiency, i.e. the *mean square error*, associated with an estimator. If the estimator S^2 of the parameter σ_X^2 is considered, the mean square error is given by:

$$E\left[\left(S^2 - \sigma_X^2\right)^2\right] \tag{5.65}$$

The efficiency of the estimator S^2 can be shown to be better than the efficiency of the estimator S^2_{unbiased} and then the choice stands between a less efficient but *unbiased estimator* or a more efficient but *biased estimator*. A number of other criteria such as *invariance*, *consistency*, *sufficiency* and *robustness* may be considered when comparing estimators. These characteristics will not be considered here but it is simply noted that the maximum likelihood method estimators in general have equally good or better characteristics than any other estimators. For more details the reader is referred to Benjamin and Cornell [4] where further references to specialized literature are also provided.

5.7.3 Confidence Intervals

As seen in the previous section, estimators are associated with *statistical uncertainty* and thus it is essential that this uncertainty is quantified and taken into account in the considered problem context. A classical approach for the quantification and the communication of this uncertainty is by the use of *confidence intervals*. The $1-\alpha$ confidence interval on an estimate defines an interval within which the estimated parameter will occur with a predefined probability, with α being the *significance level* (see Fig. 5.12).

If the case is considered where the standard deviation σ_X of an uncertain variable X is known with certainty and the mean value is unknown, then the double sided and symmetrical $1-\alpha$ confidence interval on the mean value is given by:

$$P\left[-k_{\alpha/2} < \frac{\bar{X}-\mu_X}{\sigma_X\frac{1}{\sqrt{n}}} < k_{\alpha/2}\right] = 1-\alpha \tag{5.66}$$

where n is the number of samples planned for the estimation of the mean value.

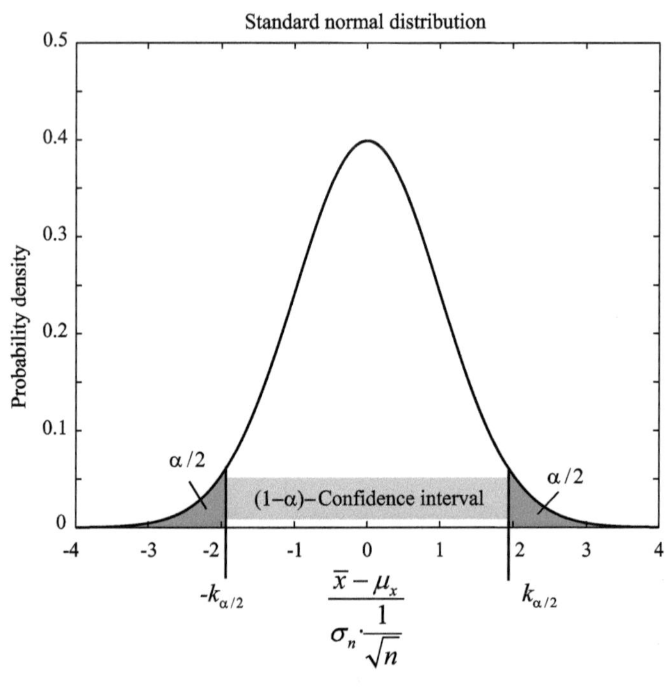

Fig. 5.12 Double sided and symmetrical $1-\alpha$ confidence interval on the mean value

Considering the case of a Normal distributed yield stress of mild construction steel, and assuming that the standard deviation of the yield stress is known to be 20 MPa and the mean value is unknown, the 0.95 confidence interval of the mean value is given by:

$$P\left[-1.96 < \frac{\bar{X} - \mu_X}{20\frac{1}{\sqrt{n}}} < 1.96\right] = 1 - 0.05 \tag{5.67}$$

where -1.96 and 1.96 are the simple lower 2.5 and upper 2.5 percentile values of the standard Normal cumulative distribution function (see also Table C.1 and Appendix B.1) and determined by:

$$k_{\alpha/2} = \Phi^{-1}\left(1 - \frac{\alpha}{2}\right) \tag{5.68}$$

where $\Phi^{-1}(\cdot)$ is the inverse standard Normal cumulative distribution function.

Assuming that 16 experiments are planned, Eq. 5.67 yields:

$$P[-9.8 < \bar{X} - \mu_X + < 9.8] = 0.95 \tag{5.69}$$

A 95% confidence interval for the true mean μ_X can then be derived with a given sample mean. Using Eq. 5.69, a confidence interval for the sample mean can also be established based on a hypothesis for the true mean, which simply states that with a probability of 0.95, the sample average of the steel yield stress will lie within an interval of ± 9.8 MPa of the true mean value.

From Eq. 5.66 it is seen that the confidence interval limits depend on α, n and σ_X. Typically α is chosen as 0.1, 0.05 and 0.01 in engineering applications. Narrow confidence intervals may be achieved by increasing the number of experiments, which on the other hand, may be expensive to achieve and in some cases not even possible for practical reasons.

5.8 Testing for Statistical Significance

In practical engineering problems, the engineer is often confronted with the challenge of deriving simple operational conclusions based on an often small set of data exhibiting a high degree of variability. An example of such a situation is the geotechnical engineer attempting, by means of a limited number of "on-site" vane tests, to verify that an empirical soil strength model based on soil specimen laboratory tests is unbiased. Another example is the materials expert pursuing to verify, by analysis of the chloride content of drilled concrete cylinders samples, that the mean value of the surface concentration of chlorides on a concrete structure can be assumed equal to the value assumed in the design basis for the structure. Yet another kind of problem is the selection and/or verification of probabilistic models as shall be seen later in Sect. 5.9.

It is essential that the basis for conclusions in problems such as those outlined above is made consistently from case to case (and from engineer to engineer), and that the variability of the observed data and the amount of data is taken appropriately into account. One approach which facilitates the support of such conclusions is the formulation and testing of hypothesis—*hypothesis testing*, which will be introduced in the following section.

Consider the example concerning the surface concentration of chlorides on a concrete structure. In the design basis for the structure, it was assumed that the surface concentration of chlorides (measured in percentage of total concrete weight) would be 0.3%. Suppose now that the materials expert has studied the chloride contents of concrete cylinders taken from 10 different locations of the considered structure. Even though the materials expert has collected a data set of 10 surface concentration values, the observed mean value, also called the sample average $\bar{X}$ (in general terms one of several possible *sample statistics*, i.e. functions of the tested or otherwise observed data), will in some cases be below and in some cases above the true mean μ_X of the surface chloride concentration. The question is if it can be concluded, on the basis of the observed statistic, that the sample average deviates statistically significantly from the assumed mean value. To arrive at the solution of such problems hypothesis testing includes *operating rules* that describe how to reach a conclusion, which provides a means for assessing the percentage α of times where the reached conclusions are wrong in one way or the other.

5.8.1 The Hypothesis Testing Procedure

Continuing with the example introduced in the foregoing, a first step is the formulation of the *null-hypothesis* H_0, i.e. expressing that the true mean value μ_X of the surface chloride concentration is equal to the assumed value of 0.3%. The next step is to formulate an *operating rule* on the basis of which the null-hypothesis can be either accepted or rejected given the test results. An operating rule could be to accept the null-hypothesis H_0 if the sample average $\bar{X}$ of the surface chloride concentration is within the interval 0.3% $\pm\Delta$ or otherwise to reject it. Rejecting the null-hypothesis implies accepting the *alternate hypothesis* H_1 that the true mean value μ_X is different from the assumed value. Typically, the value Δ is selected such that the probability α if the sample average $\bar{X}$ being outside the interval given by Δ is small, say 0.1.

Two types of errors may occur, namely, rejecting the null-hypothesis H_0 when it is true or accepting it when it is false. These two different types of errors are referred to as *Type I and Type II errors* respectively (Table 5.4). It is important to note that performing *Type I* as well as *Type II* errors may be associated with severe consequences. The selection of an appropriate value for α should reflect this. A possible approach for the selection of α is using Bayesian decision analysis. The general principles of Bayesian decision analysis are introduced in Chap. 7.

Table 5.4 Type I and Type II errors

Judgment	Truth	
	H_0 is right	H_1 is right
Acceptance of H_0	Right judgment	Type II error
Acceptance of H_1	Type I error	Right judgment

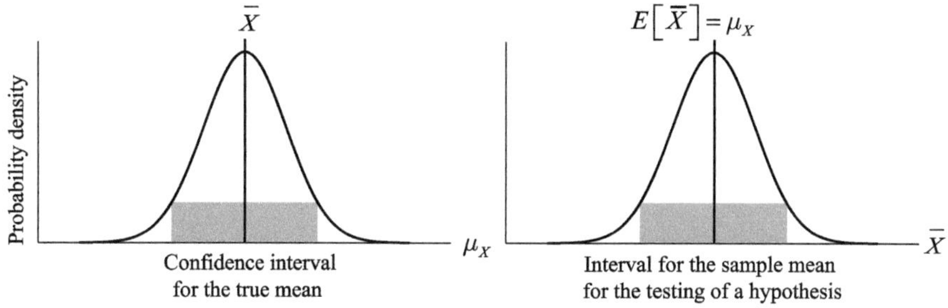

Fig. 5.13 Illustration of the difference between the confidence interval and the interval for hypothesis testing

In summary the procedure is:

- Formulate a *null-hypothesis* H_0 expressing that the desired condition is fulfilled and formulate the *alternate hypothesis* H_1. Both hypotheses should be formulated in terms of a sample statistic.
- Formulate an *operating rule* such that the formulated null-hypothesis H_0 may easily be either accepted or rejected on the basis of the observation of the sample statistic. Operating rules are typically formulated by means of a constant Δ.
- Select a significance level α for conducting the test (i.e. the probability of occurrence of a Type I error). This should be done with due consideration of the consequences of performing this type of error.
- By statistical analysis of the sample statistic, identify the value of Δ resulting in a probability α of performing a Type I error.
- Perform the planned testing, evaluate the corresponding sample statistic and check which hypothesis is supported by the experiment.
- Provided that the *null-hypothesis* H_0 is not supported by the experiment, it is classified as significant at the α -significance level and rejected; otherwise it is accepted.

Figure 5.13 shows the difference between the confidence interval and the interval for hypothesis testing. If the interest concerns the expected value of a random variable, the confidence interval indicates a range deduced by the sample where the true mean lies with a certain probability. In hypothesis testing, an interval is specified where the sample mean has to be located so that the hypothesis for the true mean of the population can be accepted.

In the following section a selection of typical cases for significance testing will be presented.

5.8.2 Testing of the Mean with Known Variance

The example of the surface concentration of chlorides on a concrete structure is considered again. Based on extensive experience obtained from the assessment of many structures, the variance of the chloride surface concentration is assumed to be equal to $\sigma_x^2 = (0.04)^2 = 0.0016$.

Let the null hypothesis H_0 be formulated as the true mean μ_x being equal to 0.3%. The alternate hypothesis H_1 is then simply given by $\mu_x \neq 0.3\%$. The considered statistic is the sample average $\bar{X}$ which may be assumed to be Normal distributed. The *operating rule* specifies that the null hypothesis H_0 should be accepted at the α-significance level if:

$$0.3 - \Delta \leq \bar{X} \leq 0.3 + \Delta \tag{5.70}$$

where Δ is determined such that the probability of $\bar{X}$ being outside the interval is equal to α i.e.:

$$P\left(0.3 - \Delta \leq \bar{X} \leq 0.3 + \Delta\right) = 1 - \alpha \tag{5.71}$$

Choosing $\alpha = 0.1$, Δ can be determined from Eq. 5.71 as (see also Appendix B.1.2):

$$\begin{aligned} &\Phi\left(\frac{\Delta}{\frac{0.04}{\sqrt{10}}}\right) - \Phi\left(\frac{-\Delta}{\frac{0.04}{\sqrt{10}}}\right) = 0.9 \Rightarrow \Phi\left(\frac{\Delta}{\frac{0.04}{\sqrt{10}}}\right) - \left(1 - \Phi\left(\frac{\Delta}{\frac{0.04}{\sqrt{10}}}\right)\right) = 0.9 \\ &\Rightarrow 2\Phi\left(\frac{\Delta}{\frac{0.04}{\sqrt{10}}}\right) - 1 = 0.9 \Rightarrow \frac{\Delta}{\frac{0.04}{\sqrt{10}}} = \Phi^{-1}\left(\frac{0.9+1}{2}\right) \\ &\Rightarrow \Delta = \frac{0.04}{\sqrt{10}} \cdot 1.645 = 0.0208 \end{aligned} \tag{5.72}$$

yielding $\Delta = 0.0208$. If the observed sample average over 10 samples lies within the interval $[0.28; 0.32]$, the null hypothesis $\mu_x = 0.3\%$ cannot be rejected at the $\alpha = 0.1$ significance level.

Assuming that investigations of the material expert resulted in the following data set $\hat{\mathbf{x}} = (0.33, 0.32, 0.25, 0.31, 0.28, 0.27, 0.29, 0.3, 0.27, 0.28)^T$ the sample average is equal to $\bar{x} = 0.29$. This is seen to be within the boundaries of the interval given by Δ and the null hypothesis H_0 cannot be rejected.

5.8.3 Some Remarks on Testing

In the foregoing, a simple case of testing of the mean with known variance has been introduced for the assessment of observed data. In the literature, several other tests such as the testing of the mean with unknown variance, the testing of the variance and the testing for comparison of two or more data sets can be found.

For all these tests, it should be noted that the different ways of formulating the null hypothesis H_0 and the different choices of the significance level α have an impact on the probability of the *Type I and Type II errors*, respectively. The optimal choice is a decision problem which can be solved by considering a proper weighing of costs and benefits. This is also reflected in the way different organizations formulate their null hypothesis H_0. An organization buying goods from a producing organization tends to postulate that the quality of the goods is below a given criterion, unless it can be shown by testing to be statistically significantly above the specified criterion. This encourages the producing organization to attempt to reduce the variance of the quality of the produced goods.

Lecture 11 (Aim of the Present Lecture) The aim of the present lecture is to address the problem of model verification and comparison. Having established a probabilistic model in terms of a probability distribution and estimated probability distribution parameters, the issue here is how to evaluate the appropriateness of the established model by means of data. Furthermore, in order to provide guidance on the comparison of the goodness of equally acceptable models, a basis for the comparison of two or more models is provided.

On the basis of the lecture it is expected that the reader will acquire knowledge and skills with regard to:

- How can probabilistic models be evaluated and validated on the basis of statistical tests?
- What is the idea behind the χ^2-goodness of fit test and how is it performed?
- How is the χ^2-goodness of fit test applied for the testing of the validity of continuous random variable models?
- What is the idea behind the Kolmogorov-Smirnov goodness of fit test and how is it performed?
- How is the Quantile-plot related to the Kolmogorov-Smirnov goodness of fit test?
- How conclusive are statistical tests for the purpose of model verification and what must be kept in mind?
- How can probabilistic models be compared with regard to appropriateness?

5.9 Model Evaluation by Statistical Testing

In Sect. 5.2.1, a highly qualitative method—the probability paper—was introduced for the identification of a family or type of probability distributions representing data obtained from observations or experimental results. This method, in conjunction with a physical understanding of the mechanism generating the observed data, represents a very pragmatic approach for establishing at least a preliminary model assumption. The next step in model building typically concerns the assessment of the parameters of the assumed distribution function and to this end the method of moments and the maximum likelihood method were introduced. From the foregoing sections it is obvious that the quality of the established model is a combination of the appropriateness of the selected probability distribution and the estimated parameters, i.e. the goodness of fit. It would thus be of significant interest to be able to assess the goodness of fit in a quantitative way thereby enabling for a systematic and consistent way of justifying or rejecting model assumptions. For this purpose, the classical statistical distribution tests, i.e. the goodness of fit tests have been developed. A number of different types of tests have been developed in the past, in part with very specialized and, consequently, limited applicability over different application areas. In the following sections, two such tests—the χ^2- and the *Kolmogorov-Smirnov* goodness of fit tests—will be explained as these have gained some importance in a broader range of engineering application areas.

In principle the χ^2-goodness of fit test is applicable only for discrete probability distributions. However, it may easily be adapted to continuous probability distributions as shall be seen in the following. The Kolmogorov-Smirnov goodness of fit test, on the other hand, is only applicable for continuous probability distributions.

5.9.1 The Chi-Square (χ^2)-Goodness of Fit Test

The χ^2-goodness of fit test is applicable for discrete cumulative distribution functions $P(x_i)$ e.g. defined by:

$$P(x_i) = \sum_{j=1}^{i-1} p(x_j) \tag{5.73}$$

Intuitively, postulating a cumulative distribution function of the type as given in Eq. 5.73, the differences between predicted frequencies $N_{p,i}$ (using the assumed model) and the observed frequencies $N_{o,i}$, should indicate the quality of the postulated cumulative distribution function; this is the idea behind the χ^2-test.

Assume that the random variable X_j is sampled n times. Then the expected value and the variance of X_j i.e. $E[X_j]$and $Var[X_j]$ are given by (see also Sect. 4.3):

$$\begin{aligned} E[X_j] &= np(x_j) = N_{p,j} \\ Var[X_j] &= np(x_j)(1 - p(x_j)) = N_{p,j}(1 - p(x_j)) \end{aligned} \tag{5.74}$$

In accordance with the *central limit theorem* and provided that the postulated model is correct, it is reasonable to assume that the standardized random deviations of the sample frequency histogram from the postulated frequency histogram ε_j i.e.:

$$\varepsilon_j = \frac{N_{o,j} - N_{p,j}}{\sqrt{N_{p,j}(1 - p(x_j))}} \tag{5.75}$$

are standard Normal distributed. This however assumes that the number of samples of each of the x_j values is large enough for the *central limit theorem* to be valid. If, however, not just the absolute values of the deviations but rather the squared deviations ε_j^2, summed up over all possible values of the discrete random variable i.e. for $j = 1, 2, \ldots, k$, are considered, it is known from Sect. 5.6.1 that this sample statistic is Chi-square distributed:

$$\varepsilon^2 = \sum_{j=1}^{k} \varepsilon_j^2 = \sum_{j=1}^{k} \frac{(N_{o,j} - N_{p,j})^2}{N_{p,j}(1 - p(x_j))} \tag{5.76}$$

Due to the fact that the numbers of realizations of the discrete random variables are dependent, the statistic given by Eq. 5.76 does not in fact have k *degrees of freedom* but only $k - 1$. Furthermore, for the same reason, each term in Eq. 5.76 shall be reduced with the factor $(1 - p(x_j))$ whereby finally the modified statistic ε_m^2 is obtained:

$$\varepsilon_m^2 = \sum_{j=1}^{k} \frac{(N_{o,j} - N_{p,j})^2}{N_{p,j}} \tag{5.77}$$

which is Chi-square distributed with $k - 1$ degrees of freedom.

Following the principles given in Sect. 5.8, it is thus possible to formulate and test, at the α-significance level, the null-hypothesis H_0 that the postulated distribution function is not in contradiction with the observed data. The operating rule states that the null hypothesis cannot be accepted if $\varepsilon_m^2 \geq \Delta$ where the critical value Δ with which the sample statistic shall be compared can be calculated from Table C.3 such as:

$$P(\varepsilon_m^2 \leq \Delta) = 1 - \alpha \tag{5.78}$$

It should be underlined that the alternate hypothesis H_1 is less informative in the sense that this hypothesis in principle envelopes all possible distributions and distribution parameters except those of the postulated probability distribution.

As an example, consider that a Normal distribution with mean value $\mu = 33$ and standard deviation $\sigma = 5$ is postulated as being representative for the data of the observed concrete compressive strengths presented in Table 5.1—this postulate is the null hypothesis H_0. It is clear that the concrete compressive strength is a continuous variable but this can be discretized by dividing the continuous sample space into intervals. The probability of a realization of the continuous random variable in each of the intervals is given as the probability that the outcome of the random variable is smaller that the upper boundary of the interval minus the probability that the outcome of the random variable is smaller than the lower boundary of the interval. It is then possible, adopting the same procedure as explained in the above, for discrete random variables to plot the histograms with the observed and predicted frequencies, $N_{o,i}$ and $N_{p,i}$, respectively, for the different data ranges $i = 1, 2, \ldots, k$ in one figure, see Fig. 5.14.

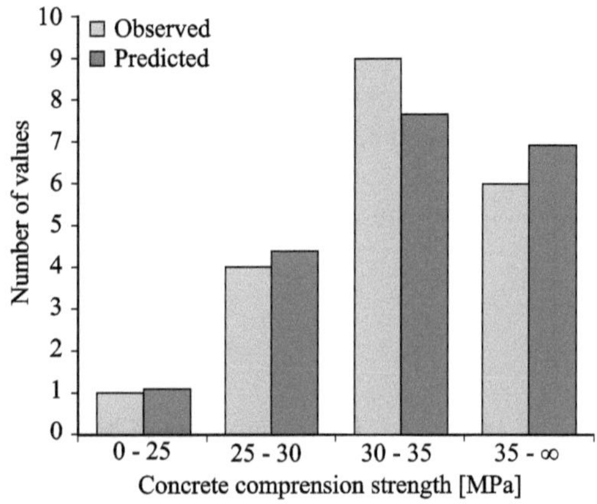

Fig. 5.14 Predicted and sample histograms for the compressive strength of concrete (data from Table 5.1)

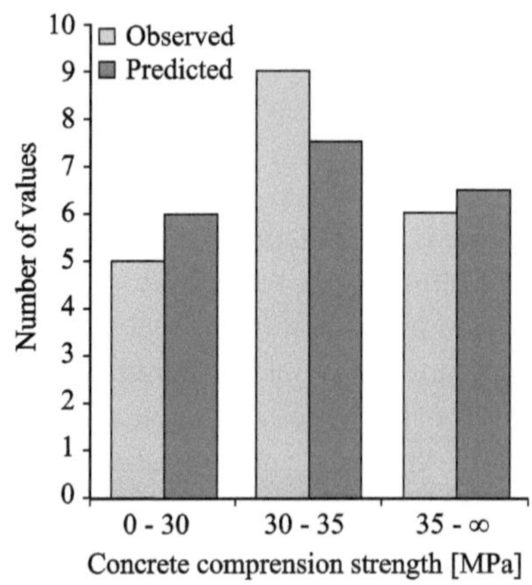

Fig. 5.15 Predicted and sample histograms for the compressive strength of concrete with lumped data for the lower interval (see Fig. 5.14)

From Fig. 5.14 it is seen that the chosen discretized implies that the number of different data ranges k is equal to 4. However, it is noted that the observed and the predicted frequencies in the lower interval is relatively small and it is doubtful if the conditions prevailing the Normal distribution assumption are fulfilled. To overcome this problem, it is recommended in the literature, see e.g. Benjamin and Cornell [4] to lump the data in adjacent intervals such that the number of observations in each interval is about 5 or larger. Lumping the frequencies in the two lower intervals yields the histograms shown in Fig. 5.15.

Following the approach outlined in the foregoing, the statistic given in Eq. 5.77 is now evaluated as summarized in Table 5.5.

Finally, in order to either reject or accept the null-hypothesis H_0, the sample statistic, as calculated in Table 5.5, must be compared with a critical value Δ given by Eq. 5.78. In the present case, if the test is performed at a 5% significance level, the value $\Delta = 5.9915$ can be calculated from a Chi-square distribution with 2 degrees of freedom. By comparison of the sample statistic in Table 5.5, i.e. 0.406829 with

Table 5.5 Calculation sheet for the χ^2-goodness of fit test of the concrete compressive strength example (raw data are given in Table 5.1)

Interval x_j	Number of observed values $N_{o,j}$	Predicted probability $p(x_j)$	Predicted number of observations $N_{p,j} = 20p(x_j)$	Sample statistic equation 5.77
0–30	5	0.274253	5.485061	0.042896
30–35	9	0.381169	7.623373	0.248591
35–∞	6	0.344578	6.891566	0.115342
			Sum	0.406829

Table 5.6 Calculation sheet for the χ^2-goodness of fit test of the concrete compressive strength example with reduced number of degree of freedom (raw data in Table 5.1)

Interval x_j	Number of observed values $N_{o,j}$	Predicted probability $p(x_j)$	Predicted number of observations $N_{p,j} = 20p(x_j)$	Sample statistic equation $\frac{(N_{o,j}-N_{p,j})^2}{N_{p,j}}$
0–30	5	0.229425	4.588507	0.036902
30–35	9	0.459861	9.197211	0.004229
35–∞	6	0.310714	6.214283	0.007389
			Sum	$\varepsilon_m^2 = 0.044852$

the critical value $\Delta = 5.9915$, it is seen that the null hypothesis cannot be rejected at the 5% significance level.

In the foregoing example not only the type of distribution, but also the parameters of the distribution were postulated. In practice, it is often the case that the parameters of the distribution are estimated first and thereafter the test for distribution type is performed. This is, in principle, possible following exactly the same approach as outlined in the above, with the modification that the number of degrees of freedom is reduced with the number of parameters estimated from the available data. If, as in the example concerning the concrete cube compressive strength, it is assumed that first the experiment data are used to assess the standard deviation of the distribution as shown in Sect. 5.6, the number of degrees of freedom is reduced to 1.

Postulating as before, a Normal distribution with mean equal to $\mu = 33$ but now with the standard deviation found in Sect. 5.3, i.e. $\sigma = 4.05$, the calculations are modified as shown in Table 5.6.

With only 1 degree of freedom the critical level Δ is reduced to $\Delta = 3.84$ but the null-hypothesis still cannot be rejected at the 5% significance level.

From the above, it is seen that the available data simply do not permit that both of the distribution parameters are first estimated and thereafter the distribution and parameters postulates tested. The number of degrees of freedom is not sufficient. In engineering applications this problem is not unusual as the available data are generally sparse. In other areas, such as in production industries, the available amount of data is generally very substantial and the merits of statistical testing are more obvious.

5.9.2 The Kolmogorov-Smirnov Goodness of Fit Test

Whereas the χ^2-goodness of fit test takes basis in a statistic quantifying the squared errors between the observed sample histogram and the predicted postulated histogram, the goodness of fit test due to Kolmogorov-Smirnov utilizes a sample statistic formulated in terms of the cumulative distributions. The χ^2-goodness of fit test can be applied in cases where both distribution and parameters are postulated, as well as in cases where only the distribution is postulated and the parameters estimated, using the same data utilized for the testing. This is not the case for the Kolmogorov-Smirnov test. Here both the distribution family and parameters must be postulated.

If the observed cumulative distribution function $F_o(\hat{x}_i^o)$ is written as:

$$F_o(\hat{x}_i^o) = \frac{i}{n} \tag{5.79}$$

where $\hat{x}_i^o$ is the ith smallest observation in the sample of size n and the postulated cumulative distribution function is $F_p(x)$, then the following statistic may be utilized for significance testing of the null hypothesis that the observed data do not deviate statistically significantly from the postulated distribution function:

$$\varepsilon_{\max} = \max_{i=1}^{n} \left[\left| F_o(\hat{x}_i^o) - F_p(\hat{x}_i^o) \right| \right] = \max_{i=1}^{n} \left[\left| \frac{i}{n} - F_p(\hat{x}_i^o) \right| \right] \tag{5.80}$$

The distribution of the statistic $\varepsilon_{\max}$ is tabulated for relevant values of α and n in Table 5.7.

The null hypothesis H_0 may be formulated expressing that the observed data follows the postulated cumulative distribution function.

The operating rule specifies that the null hypothesis H_0 cannot be accepted at the α-significance level if: $\varepsilon_{\max} \geq \Delta$ where the critical value Δ can be calculated such that:

$$P(\varepsilon_{\max} \leq \Delta) = 1 - \alpha \tag{5.81}$$

where Δ is determined from Table 5.7 (or Table C.4).

Consider again the example concerning the compressive strength of concrete. As before, it is postulated that the data is representative for a Normal distribution with mean value $\mu = 33$ and a standard deviation equal to $\sigma = 5$. By inspection of Fig. 5.16 it is seen that the largest deviation between the observed cumulative distribution function and the predicted postulated probability distribution occurs for the 18th data point corresponding to a concrete cube compressive

Table 5.7 Tabulated values of the Kolmogorov-Smirnov statistic for different significance levels α and sample size n

α	n											
	1	5	10	15	20	25	30	40	50	60	70	80
0.01	0.9950	0.6686	0.4889	0.4042	0.3524	0.3166	0.2899	0.2521	0.2260	0.2067	0.1917	0.1795
0.05	0.9750	0.5633	0.4093	0.3376	0.2941	0.2640	0.2417	0.2101	0.1884	0.1723	0.1598	0.1496
0.1	0.9500	0.5095	0.3687	0.3040	0.2647	0.2377	0.2176	0.1891	0.1696	0.1551	0.1438	0.1347
0.2	0.9000	0.4470	0.3226	0.2659	0.2315	0.2079	0.1903	0.1654	0.1484	0.1357	0.1258	0.1179

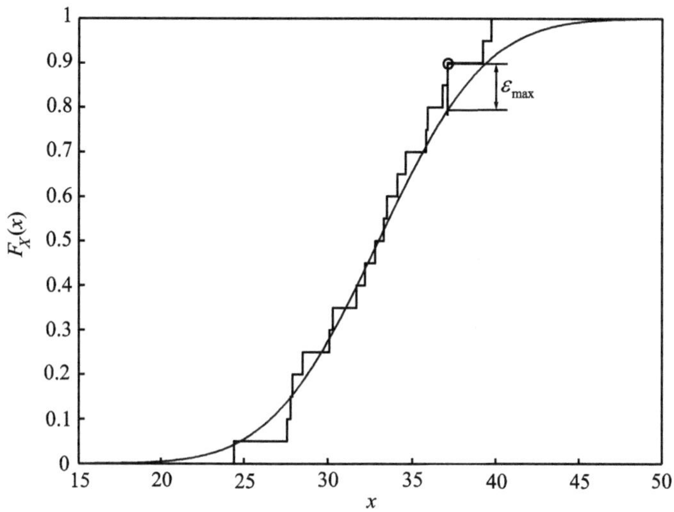

Fig. 5.16 Comparison between the observed concrete cube compressive strength data and the postulated Normal distribution

strength of 37.1 MPa. For this value the postulated cumulative distribution function yields:

$$F_p(\hat{x}_{18}^o) = \Phi\left(\frac{37.1 - 33}{5}\right) = \Phi\,(0.82) = 0.794 \tag{5.82}$$

and the observed cumulative distribution function yields:

$$F_o(\hat{x}_{18}^o) = \frac{i}{n} = \frac{18}{20} = 0.9 \tag{5.83}$$

whereby the sample statistic becomes $\varepsilon_{\max} = 0.9 - 0.794 = 0.106$. From Table 5.7 the critical value Δ for $n = 20$ and a 5% significance level is 0.29. Since the sample statistic 0.106 is smaller than the critical value 0.29, the null hypothesis cannot be rejected.

5.9.3 Model Comparison

In the foregoing sections statistical tests were introduced as means for evaluating the goodness of fit of a given postulated distribution function to observed data. These tests can, however, only be applied to assess the plausibility of a given distribution being representative for the observed data. Other postulated distribution functions could also be representative for the observed data. The question thus remains how to select between two postulated distributions which both cannot be rejected by testing as possible candidates. To this end two possibilities might be considered, namely by comparison of the *sample likelihood* defined by Eq. 5.6 or by comparison of the likelihood of the sample statistics given by Eq. 5.77 or Eq. 5.80. Direct comparison of the sample statistic for the χ^2-goodness of fit test is not a consistent means for comparison as the number of degrees of freedom may be different for the cases considered.

As an example consider the two cases where the χ^2-goodness of fit test was applied first for testing the goodness of fit for a postulated Normal distribution with postulated parameters $\mu = 33$, $\sigma = 5$ and thereafter a postulated Normal distribution for which the parameters were estimated from the data set, i.e. $\mu = 33$, $\sigma = 4.05$. For the first case the sample statistic is equal to 0.4068 and the number of degrees of freedom is 2. For the second case the sample statistic is equal to 0.44852 and the number of degrees of freedom is equal to 1.

The sample likelihood (see Eq. 5.6) of the first case ($\sigma = 5$) is $L(\theta \,|\, \hat{\mathbf{x}}) = 2.74 \cdot 10^{-23}$ and smaller than that of the second $L(\theta \,|\, \hat{\mathbf{x}}) = 1.24 \cdot 10^{-21}$. This shows that the postulate made in the second case is more probable than the postulate made in the first case.

Alternatively, the corresponding likelihood values of the density function can be compared. This can be calculated with Eq. 5.53:

$$
\begin{aligned}
f_{Y_n}(y_n) &= \frac{y_n^{(n/2-1)}}{2^{n/2}\Gamma(n/2)} \exp(-y_n/2) \\
&= \frac{0.4068^{(2/2-1)}}{2^{2/2}\Gamma(2/2)} \exp(-0.4068/2) = 0.408 \\
f_{Y_n}(y_n) &= \frac{y_n^{(n/2-1)}}{2^{n/2}\Gamma(n/2)} \exp(-y_n/2) \\
&= \frac{0.044852^{(1/2-1)}}{2^{1/2}\Gamma(1/2)} \exp(-0.044852/2) = 1.842
\end{aligned}
$$

Based on the obtained likelihood values of 0.408 and 1.842, it can be said that the postulation made in the second case is more probable than the postulate made in the first case.

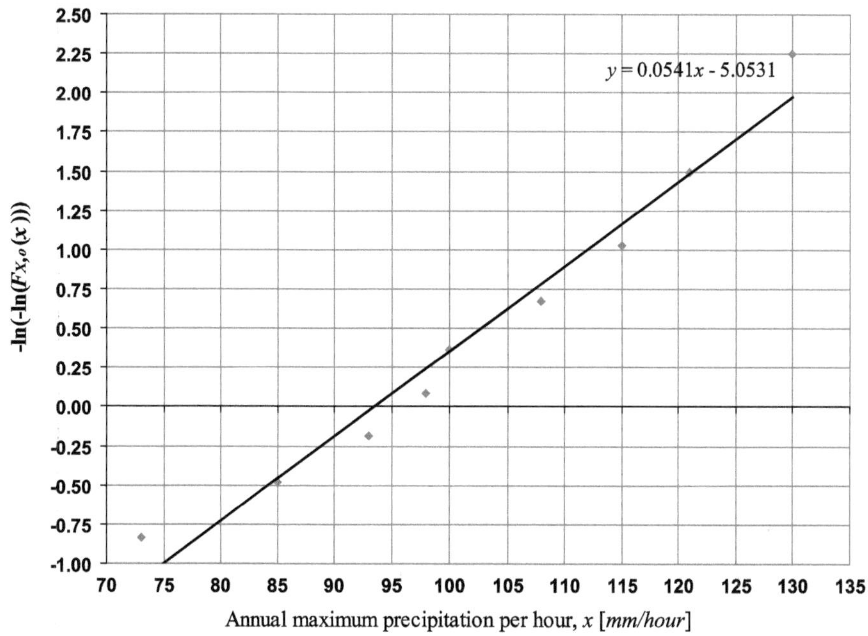

Fig. 5.17 Probability paper for the Gumbel distribution

5.10 Self Assessment Questions/Exercises

1. In Fig. 5.17, data of the annual observed maximum values of precipitation per hour (rainfall) are plotted on a probability paper for the Gumbel distribution. The "best-fit" line is also shown. Can an engineer accept the Gumbel distribution as being suitable for the modeling of the annual maximum precipitation per hour?
2. The Maximum Likelihood Method (MLM) enables engineers to calculate the distribution parameters of a random variable on the basis of data. Which of the following statement(s) is (are) correct?
   The MLM provides point estimates of the distribution parameters. ☐
   The MLM provides information about the uncertainty associated with the estimated parameters. ☐
   The MLM provides no information about the uncertainty associated with the estimated parameters. ☐
3. From past experience it is known that the shear strength of soil can be described by a Lognormal distribution. 15 samples of soil are taken from a site and an engineer wants to use the data in order to estimate the parameters of the Lognormal distribution. The engineer:
   May use a probability paper to estimate the parameters of the Lognormal distribution. ☐
   May use the maximum likelihood method to estimate the parameters of the Lognormal distribution. ☐

May use the method of moments to estimate the parameters of the Lognormal distribution. ☐

None of the above. ☐

4. Which are the steps involved in establishing a probabilistic model?
5. Express in words the following mathematical expression, where $\bar{X}$ is the sample average of a random variable X and μ_X is the true mean of the random variable:

$$P[\mu_X - 9.8 < \bar{X} < \mu_X + 9.8] = 0.95$$

6. What are the main steps of hypothesis testing?
7. An engineer wants to test the null hypothesis that the mean value of the concrete cover depth of a concrete structure corresponds to design assumptions. In a preliminary assessment, a limited number of measurements of the concrete cover depth are made, and after performing the hypothesis test the engineer accepts the null hypothesis. After a few years, a comprehensive survey of the concrete cover depth is carried out and several measurements are made. The survey shows that the mean value of the concrete cover depth does not fulfill the design assumptions. Which of the following statement(s) is(are) correct?

   In the preliminary survey the engineer has performed a Type I error. ☐

   In the preliminary survey the engineer has performed a Type II error. ☐

   In the preliminary survey the engineer has performed a Type I and a Type II error. ☐
8. Describe in a few words the significance of the use of probability paper in model selection and how it can be constructed.
9. In order to perform a χ^2 test, how does the data need to be divided?
10. What are the main differences between the χ^2 and the Kolmogorov-Smirnov goodness of fit tests?
11. Based on experience, it is known that the concrete compressive strength may be modeled by a Normal distributed random variable X with mean value $\mu_X = 30$ MPa and standard deviation $\sigma_X = 5$ MPa. The compressive strengths of 20 concrete cylinders are measured. An engineer wants to test the null hypothesis H_0 that X follows a Normal distribution with the above given parameters. He/she carries out a χ^2 goodness of fit test by dividing the samples into 3 intervals. He/she calculates a Chi-square sample statistic equal to $\varepsilon_m^2 = 0.41$. Can the engineer accept the null hypothesis at the 5% significance level?
12. An engineer wants to examine and compare the suitability of two distribution function model alternatives for a random material property. Measurements of the material property are taken. The engineer uses the two model alternatives to calculate the Chi-square sample statistics and the corresponding sample likelihoods. The results are given in Table 5.8.

    Which of the following statement(s) is(are) correct?

    The engineer may accept model 1 at the 5% significance level. ☐

Table 5.8 Statistical model characteristics

Model	Degrees of freedom	Chi-square sample statistic	Sample likelihood
1	2	0.410	0.815
2	1	0.407	0.524

The engineer may accept model 2 at the 5% significance level. ☐
Model 1 is more suitable than model 2. ☐
None of the above. ☐

Chapter 6
Methods of Structural Reliability

Lecture 12 (Aim of the Present Lecture) The aim of the present lecture is to introduce the basic theory and tools facilitating the representation of events in terms of random variables and to calculate the probability of occurrence of such events. The classical error accumulation law is introduced first and this is then set in perspective to more modern and more general tools to assess probabilities.
On the basis of the lecture it is expected that the reader will acquire knowledge on the following issues:

- How can events be represented in terms of basic random variables?
- What is a limit state function and what is a safety margin?
- What is the meaning of a reliability index and how does it relate to a failure probability?
- How to calculate the reliability index for a linear safety margin when all basic random variables are Normal distributed?
- How to calculate the reliability index in the case of nonlinear safety margins?
- What is the idea behind the Monte Carlo method?
- What are the steps in the Monte Carlo method and how are they performed?

6.1 Introduction

The first developments of *First Order Reliability Methods*, also known as FORM methods, took place in the 1960s. Since then the methods have been refined and extended significantly and by now they form one of the most important methods for reliability evaluations in structural reliability theory. Several commercial computer codes have been developed for FORM analysis and the methods are widely used in practical engineering problems and for code calibration purposes.

M.H. Faber, *Statistics and Probability Theory*,
Topics in Safety, Risk, Reliability and Quality 18,
DOI 10.1007/978-94-007-4056-3_6,

In the present chapter first the basic idea behind FORM methods is highlighted and thereafter the individual steps of the methods are explained in detail.

Finally, the basic concepts of *Monte Carlo methods* are outlined.

6.2 Failure Events and Basic Random Variables

In *reliability analysis* of technical systems and components the main problem is to evaluate the probability of failure corresponding to a specified reference period. However, other non-failure states of the considered component or system may also be of interest, such as excessive damage, unavailability, etc.

In general, any state which may be associated with consequences in terms of costs, loss of lives and impact to the environment is of interest. In the following section no differentiation will be made between these different types of states. For simplicity, all these events will be referred to as being *failure events*, bearing in mind, however, that also *non-failure states* may be considered in the same manner.

It is convenient to describe failure events in terms of functional relations. If they are fulfilled, the considered event will occur. A *failure event* may be described by a functional relation, the *limit state function* $g(\mathbf{x})$, in the following way:

$$F = \{g(\mathbf{x}) \leq 0\} \tag{6.1}$$

where the components of the vector $\mathbf{x}$ are realizations of the *basic random variables* $\mathbf{X}$ representing all the relevant uncertainties influencing the *probability of failure*. In Eq. 6.1 the failure event F is simply defined as the set of realizations of the function $g(\mathbf{x})$, which is zero or negative.

As already mentioned, events other than failure may also be of interest. In e.g. reliability updating problems, events of the following form are highly relevant:

$$I = \{h(\mathbf{x}) = 0\} \tag{6.2}$$

where I is an inspection event and $h(\mathbf{x})$ refers to the basic uncertain variables representing the inspection event.

Having defined the failure event, the probability of failure P_F may be determined by the following integral:

$$P_F = \int_{g(\mathbf{x}) \leq 0} f_{\mathbf{X}}(\mathbf{x})\, d\mathbf{x} \tag{6.3}$$

where $f_{\mathbf{X}}(\mathbf{x})$ is the joint probability density function of the random variables $\mathbf{X}$. This integral is, however, non-trivial to solve and numerical approximations are expedient. Various methods for the solution of the integral in Eq. 6.3 have been proposed including numerical integration techniques, Monte Carlo simulation and asymptotic *Laplace expansions*. Numerical integration techniques very rapidly became inefficient for increasing size of the vector $\mathbf{X}$ and are in general irrelevant. In the following section the focus is directed toward the widely applied and quite efficient FORM methods, which furthermore are consistent with the solutions obtained by asymptotic Laplace integral expansions.

6.3 Linear Limit State Functions and Normal Distributed Variables

For illustrative purposes first the case where the limit state function $g(\mathbf{x})$ is a linear function of the basic random variables $\mathbf{X}$ is considered. Then the limit state function may be written as:

$$g(x) = a_0 + \sum_{i=1}^{n} a_i x_i \tag{6.4}$$

The safety margin is then defined as $M = g(X)$. Failure can be defined by:

$$M \leq 0 \tag{6.5}$$

If the basic random variables are Normal distributed, the linear *safety margin M* defined through

$$M = a_0 + \sum_{i=1}^{n} a_i X_i \tag{6.6}$$

is also Normal distributed with mean value and variance:

$$\begin{aligned} \mu_M &= a_0 + \sum_{i=1}^{n} a_i \mu_{X_i} \\ \sigma_M^2 &= \sum_{i=1}^{n} a_i^2 \sigma_{X_i}^2 + 2\left(\sum_{i=1}^{n} \sum_{j=i+1}^{n} a_i a_j C_{X_i X_j}\right) \\ &= \sum_{i=1}^{n} a_i^2 \sigma_{X_i}^2 + 2\left(\sum_{i=1}^{n} \sum_{j=i+1}^{n} a_i a_j \rho_{ij} \sigma_{X_i} \sigma_{X_j}\right) \end{aligned} \tag{6.7}$$

ρ_{ij} are the correlation coefficients between the variables X_i and X_j.

Defining the failure event by Eq. 6.1, the probability of failure can be written as:

$$P_F = P(g(\mathbf{X}) \leq 0) = P(M \leq 0) \tag{6.8}$$

which in this simple case reduces to the evaluation of the standard Normal distribution function:

$$P_F = \Phi\left(\frac{0 - \mu_M}{\sigma_M}\right) = \Phi(-\beta) \tag{6.9}$$

where the *reliability index* β (Cornell [6] and Basler [3]) is given as:

$$\beta = \frac{\mu_M}{\sigma_M} \tag{6.10}$$

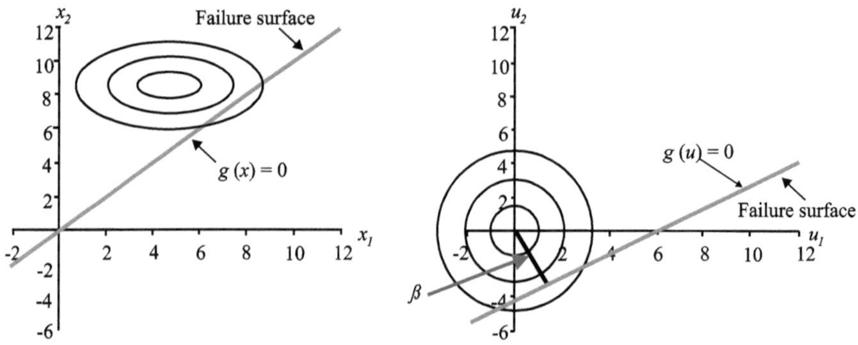

Fig. 6.1 Illustration of the two-dimensional case of a linear limit state function and Normal distributed variables **X**

In the case where the basic random variables are uncorrelated, the reliability index β as defined in Eq. 6.10 has a geometrical interpretation as illustrated in Fig. 6.1 where a two dimensional case is considered.

In Fig. 6.1 the limit state function $g(\mathbf{x})$ has been transformed into the limit state function $g(\mathbf{u})$ by standardization of the random variables as:

$$U_i = \frac{X_i - \mu_{X_i}}{\sigma_{X_i}} \tag{6.11}$$

such that the random variables U_i have mean values equal to zero and standard deviation values equal to one.

The reliability index β has the simple geometrical interpretation as the smallest distance from the line (or generally the hyper-plane) forming the boundary between the *safe domain* and the *failure domain*, i.e. the domain defined by the failure event. It should be noted that this definition of the reliability index due to Hasofer and Lind [7] does not depend on the limit state function but rather the boundary between the safe domain and the failure domain. The point on the *failure surface* with the smallest distance to the origin is commonly called the design point or the most likely failure point.

It is seen that the evaluation of the probability of failure in this simple case reduces to a simple evaluation in terms of mean values and standard deviations of the basic random variables, i.e. the first and second order information.

6.3.1 Example 6.1—Reliability of a Steel Rod—Linear Safety Margin

Consider a steel rod under pure tension loading. The rod will fail if the applied stresses on the rod cross-sectional area ($a = 10\,\text{mm}^2$) exceed the steel yield strength. The yield strength r of the rod and the annual maximum stress in the rod s are assumed to be uncertain, modeled by uncorrelated Normal distributed variables R and S, respectively. The mean values and the standard deviations of the yield strength and the loading force are given as $\mu_R = 350$ MPa, $\sigma_R = 35$ MPa and $\mu_S = 1500$ N, $\sigma_S = 300$ N respectively.

The limit state function describing the event of failure may be written as:

$$g(\mathbf{x}) = ar - s$$

whereby the safety margin M may be written as:

$$M = aR - S$$

The mean value and standard deviation of the safety margin M are thus:

$$\mu_M = 10 \cdot 350 - 2000 = 2000 \text{ N}$$
$$\sigma_M = \sqrt{10^2 \cdot 35^2 + 300^2} \text{ N}$$

whereby the reliability index may be calculated as:

$$\beta = \frac{2000}{461} = 4.33$$

Finally the annual *failure probability* is determined as:

$$P_F = \Phi(-\beta) = \Phi(-4.33) = 7.5 \cdot 10^{-6} \ [a^{-1}]$$

6.4 The Error Propagation Law

The results given in Eq. 6.7 have been applied to study the statistical characteristics of errors ε accumulating in accordance with some differentiable function $h(\mathbf{x})$, i.e.:

$$\varepsilon = h(\mathbf{x}) \tag{6.12}$$

where $\mathbf{x} = (x_1, x_2, \ldots, x_n)^T$ is a vector of realizations of the basic random variables $\mathbf{X}$ representing measurement uncertainties with mean values $\mu_{\mathbf{X}} = (\mu_{X_1}, \mu_{X_2}, \ldots, \mu_{X_n})^T$ and covariances $C_{X_i X_j} = \rho_{ij}\sigma_{X_i}\sigma_{X_j}$ where σ_{X_i} are the standard deviations and ρ_{ij} the correlation coefficients. The idea is to approximate the function $h(\mathbf{x})$ by its Taylor expansion including only the linear terms, i.e.:

$$\varepsilon \cong h(\mathbf{x}_0) + \sum_{i=1}^{n} (x_i - x_{i,0}) \left. \frac{\partial h(\mathbf{x})}{\partial x_i} \right|_{\mathbf{x}=\mathbf{x}_0} \tag{6.13}$$

where $\mathbf{x}_0 = (x_{1,0}, x_{2,0}, \ldots, x_{n,0})^T$ is the point in which the linearization is performed, normally chosen as the mean value point.

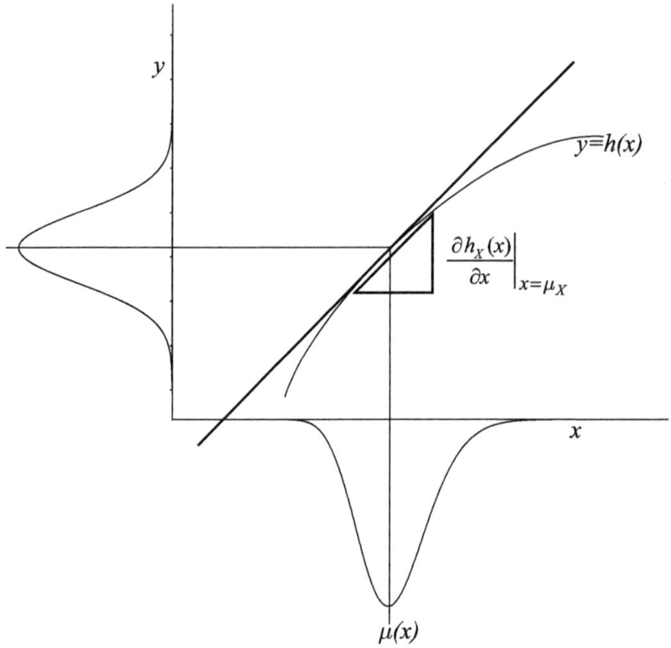

Fig. 6.2 Illustration of the error propagation law: The transformation of the density function $f_Y(y)$ according to the relation $y = g(x)$ and the linear approximation of the relation between the two random variables

$\left.\frac{\partial h(\mathbf{x})}{\partial x_i}\right|_{\mathbf{x}=\mathbf{x}_0}$, $i = 1, 2, \ldots, n$ are the first order partial derivatives of $h(\mathbf{x})$ taken in $\mathbf{x} = \mathbf{x}_0$. From Eqs. 6.13 and 6.7 it is seen that the expected value of the error $E[\varepsilon]$ can be assessed by:

$$E[\varepsilon] \cong h(\mu_{\mathbf{X}}) \tag{6.14}$$

and its variance $Var[\varepsilon]$ can be determined by:

$$Var[\varepsilon] \cong \sum_{i=1}^{n} \left(\left.\frac{\partial h(\mathbf{x})}{\partial x_i}\right|_{\mathbf{x}=\mathbf{x}_0} \right)^2 \sigma_{X_i}^2 + 2\sum_{i=1}^{n} \sum_{j=i+1}^{n} \left(\left.\frac{\partial h(\mathbf{x})}{\partial x_i}\right|_{\mathbf{x}=\mathbf{x}_0} \right) \left(\left.\frac{\partial h(\mathbf{x})}{\partial x_j}\right|_{\mathbf{x}=\mathbf{x}_0} \right) \rho_{ij}\sigma_{X_i}\sigma_{X_j} \tag{6.15}$$

It is important to notice that the variance of the error as given by Eq. 6.15 depends on the linearization point, i.e. $\mathbf{x}_0 = (x_{1,0}, x_{2,0}, \ldots, x_{n,0})^T$.

In Fig. 6.2 the basic principle of the error propagation law is represented for the (two-dimensional and therefore still representable) case that a random variable

Y depends on only one random variable X: $y = g(x)$. The function $y = g(x)$ is linearized in the mean value point and the relationship between Y and X is approximated with $y = g(\mu_X) + \frac{\partial g(x)}{\partial x}|_{X=\mu_X}(X - \mu_X)$.

The distribution of the derived random variable y represented on the vertical axis in Fig. 6.2 illustrates the approximated distribution resulting from the error propagation law. Depending on the form of the transformation function $h(x)$, the distribution can differ from the true distribution for y significantly.

6.4.1 Example 6.2—Error Propagation Law

As an example of the use of the error propagation law consider a right angle triangle ABC, where C is the right angle. The lengths of the 2 catheti a and b are measured. Due to measurement uncertainty a and b are modeled as independent Normal distributed random variables with expected values $\mu_a = 12.2$, $\mu_b = 5.1$ and standard deviations $\sigma_a = 0.4$ and $\sigma_b = 0.3$, respectively. It is assumed that a critical condition will occur if the hypotenuse c is larger than 13.5 and the probability that this condition should happen is to be assessed.

Based on the probabilistic model of a and b, the statistical characteristics of the hypotenuse c given by:

$$c = \sqrt{a^2 + b^2}$$

may be assessed through the error propagation model given by Eqs. 6.14–6.15, yielding (see also Appendix B):

$$E[c] = \sqrt{\mu_a^2 + \mu_b^2}$$

$$Var[c] = \sum_{i=1}^{n}\left(\left.\frac{\partial h(\mathbf{x})}{\partial x_i}\right|_{\mathbf{x}=\mathbf{x}_0}\right)^2 \sigma_{X_i}^2 = \frac{\mu_a^2}{\mu_a^2 + \mu_b^2}\sigma_a^2 + \frac{\mu_b^2}{\mu_a^2 + \mu_b^2}\sigma_b^2$$

which by inserting for a and b their expected values yield:

$$E[c] = \sqrt{12.2^2 + 5.1^2} = 13.22 \text{ [km]}$$

$$Var[c] = \frac{12.2^2}{12.2^2 + 5.1^2}0.4^2 + \frac{5.1^2}{12.2^2 + 5.1^2}0.3^2 = 0.15 \text{ [km}^2\text{]}$$

As seen from the above, the variance of the hypotenuse c depends on the chosen linearization point. If, instead of the mean value point, a value corresponding to the mean value plus two standard deviations was chosen, the variance of c would have been:

$$
\begin{aligned}
Var[c] &= \sum_{i=1}^{n} \left(\left. \frac{\partial h(\mathbf{x})}{\partial x_i} \right|_{\mathbf{x}=\mathbf{x}_0} \right)^2 \sigma_{X_i}^2 \\
&= \frac{(\mu_a + 2\sigma_a)^2}{(\mu_a + 2\sigma_a)^2 + (\mu_b + 2\sigma_b)^2} \sigma_a^2 + \frac{(\mu_b + 2\sigma_b)^2}{(\mu_a + 2\sigma_a)^2 + (\mu_b + 2\sigma_b)^2} \sigma_b^2 \\
&= \frac{13^2}{13^2 + 5.7^2} 0.4^2 + \frac{5.7^2}{13^2 + 5.7^2} 0.3^2 = 0.149
\end{aligned}
$$

which can be shown to imply a 0.3% reduction of the probability that the hypotenuse is larger than 13.5. Even though such a change seems small, it could be of importance in a practical situation where the consequences of errors can be significant.

6.5 Non-linear Limit State Functions

When the limit state function is non-linear in the basic random variables $\mathbf{X}$, the situation is not as simple as outlined in the previous section. An obvious approach is, however, considering the error propagation law explained in the foregoing, to represent the failure domain in terms of a linearization of the boundary between the *safe domain* and the *failure domain*, i.e. the failure surface. Still, the question remains how to do this appropriately.

Hasofer and Lind [7] suggest performing this *linearization* in the *design point* of the failure surface represented in normalized space. The situation is illustrated in the 2-dimensional space in Fig. 6.3.

In Fig. 6.3 a principal sketch is given, illustrating that the failure surface is linearized in the design point $\mathbf{u}^*$ by the line $g'(\mathbf{u}) = 0$. The $\boldsymbol{\alpha}$-vector is the outward directed Normal vector to the failure surface in the design point $\mathbf{u}^*$ i.e. the point on the linearized failure surface with the shortest distance β to the origin.

As the limit state function is in general non-linear, one does not know the design point in advance. This has to be found e.g. by solving the following optimization problem:

$$
\beta = \min_{\mathbf{u} \in \{g(\mathbf{u})=0\}} \sqrt{\sum_{i=1}^{n} u_i^2} \tag{6.16}
$$

This problem may be solved in a number of different ways. Provided that the limit state function is differentiable, the following iteration scheme may be followed:

$$
\alpha_i = \frac{-\frac{\partial g}{\partial u_i}(\beta \boldsymbol{\alpha})}{\left[\sum_{i=1}^{n} \left(\frac{\partial g}{\partial u_i}(\beta \boldsymbol{\alpha}) \right)^2 \right]^{1/2}}, \quad i = 1, 2, \ldots, n \tag{6.17}
$$

$$
g(\beta\alpha_1, \beta\alpha_2, \ldots, \beta\alpha_n) = 0 \tag{6.18}
$$

First, a design point is guessed $\mathbf{u}^* = \beta\boldsymbol{\alpha}$ and inserted into Eq. 6.17 whereby a new Normal vector $\boldsymbol{\alpha}$ to the failure surface is achieved. Then this $\boldsymbol{\alpha}$-vector is inserted into Eq. 6.18, from which a new β-value is calculated.

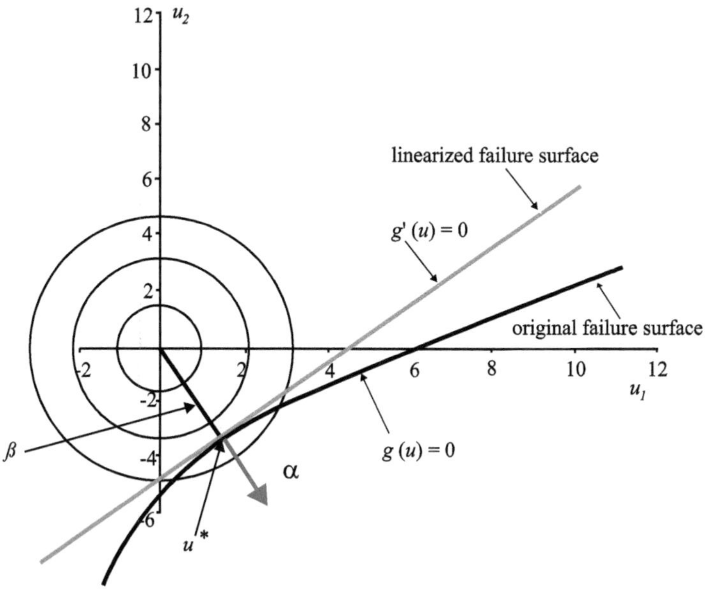

Fig. 6.3 Illustration of the linearization proposed by Hasofer and Lind [7] in standard Normal space

The iteration scheme will converge in a few (say normally 6–10) iterations and provides the design point $\mathbf{u}^*$ as well as the reliability index β and the outward normal to the failure surface in the design point $\boldsymbol{\alpha}$. As already mentioned, the reliability index β may be related directly to the probability of failure. The components of the $\boldsymbol{\alpha}$-vector may be interpreted as sensitivity factors giving the relative importance of the individual random variables for the reliability index β.

Second Order Reliability Methods (SORM) follow the same principles as FORM. However, as a logical extension of FORM, the failure surface is expanded to the second order in the design point. The result of a SORM analysis may be given as the FORM β multiplied with a correction factor evaluated on the basis of the second order partial derivatives of the failure surface in the design point. The SORM analysis becomes exact for failure surfaces given as a second order polynomial of the basic random variables. In general, however, the result of a SORM analysis can be shown to be asymptotically exact for any shape of the failure surface as β approaches infinity. The interested reader is referred to the literature for the details of SORM analysis, e.g. Madsen et al. [9].

6.5.1 Example 6.3—FORM—Non-linear Limit State Function

Consider again the steel rod from Example 6.1. However, now it is assumed that the cross sectional area of the steel rod A is also uncertain.

The steel yield strength R is Normal distributed with mean values and standard deviation $\mu_R = 350$ MPa, $\sigma_R = 3$ MPa5 and the loading S is Normal distributed with mean value and standard deviation $\mu_S = 1500$ N, $\sigma_S = 300$ N. Finally, the cross sectional area A is assumed Normal distributed with mean value and standard deviation $\mu_A = 10$ mm^2 $\sigma_A = 1$ mm^2.

The limit state function may be written as:

$$g(\mathbf{x}) = ra - s$$

Now the first step is to transform the Normal distributed random variables R, A and S into standardized Normal distributed random variables, i.e.:

$$U_R = \frac{R - \mu_R}{\sigma_R}$$

$$U_A = \frac{A - \mu_A}{\sigma_A}$$

$$U_S = \frac{S - \mu_S}{\sigma_S}$$

The limit state function may now be written in the space of the standardized Normal distributed random variables as:

$$\begin{aligned} g(u) &= (u_R\sigma_R + \mu_R)(u_A\sigma_A + \mu_A) - (u_S\sigma_S + \mu_S) \\ &= (35u_R + 350)(1u_A + 10) - (300u_S + 1500) \\ &= 350\mathrm{u_R} + 350u_A - 300u_S + 35u_Ru_A + 2000 \end{aligned}$$

The reliability index and the design point may be determined in accordance with Eqs. 6.17 and 6.18 as (see also Appendix B):

$$\beta = \frac{-2000}{350\alpha_R + 350\alpha_A - 300\alpha_S + 35\beta\alpha_R\alpha_A}$$

$$\alpha_R = -\frac{1}{k}(350 + 35\beta\alpha_A)$$

$$\alpha_A = -\frac{1}{k}(350 + 35\beta\alpha_R)$$

$$\alpha_S = \frac{300}{k}$$

with:

$$k = \sqrt{(350 + 35\beta\alpha_A)^2 + (350 + 35\beta\alpha_R)^2 + (300)^2}$$

which by calculation gives the iteration history shown in Table 6.1.

From Table 6.1 it is seen that the basic random variable S modeling the load on the steel rod is slightly dominating with an α-value equal to 0.6087. Furthermore it

Table 6.1 Iteration history for the non-linear limit state example

Iteration	Start	1	2	3	4	5
β	3.0000	3.6719	3.7392	3.7444	3.7448	3.7448
α_R	−0.5800	−0.5602	−0.5612	−0.5610	−0.5610	−0.5610
α_A	−0.5800	−0.5602	−0.5612	−0.5610	−0.5610	−0.5610
α_S	0.5800	0.6101	0.6085	0.6088	0.6087	0.6087

is seen that both the variables R and A are acting as resistance variables as their α-values are negative. The annual failure probability for the steel rod is determined as:

$$P_F = \Phi(-\beta) = \Phi(-3.7448) = 9.02 \cdot 10^{-5}\ [a^{-1}]$$

6.6 Simulation Methods

Consider again the probability integral defined in Eq. 6.3:

$$P_f = \int_{g(\mathbf{x})\leq 0} f_{\mathbf{X}}(\mathbf{x})\, d\mathbf{x} \tag{6.19}$$

It has been seen that FORM methods may successfully be applied for the evaluation of this integral. The integral may also be estimated by so-called *simulation techniques*. In the literature, a large variety of simulation techniques may be found and a treatment of these will not be given in the present text. Here it is just noted that simulation techniques have proven their value especially for problems where the representation of the limit state function is associated with difficulties. Such cases are e.g. when the limit state function is not differentiable or when several design points contribute to the failure probability.

However, as all simulation techniques have their origin in the so-called crude *Monte Carlo Method*, the principles of this simulation technique will be briefly outlined below.

The basis for simulation techniques is well illustrated by rewriting the probability integral in Eq. 6.19 by means of an *indicator function*:

$$P_F = \int_{g(\mathbf{x})\leq 0} f_{\mathbf{X}}(\mathbf{x})\, d\mathbf{x} = \int I\left[g(\mathbf{x}) \leq 0\right] f_{\mathbf{X}}(\mathbf{x})\, d\mathbf{x} \tag{6.20}$$

where the integration domain is changed from the part of the sample space of the vector $\mathbf{X} = (X_1, X_2, \ldots, X_n)^T$ for which $g(\mathbf{x}) \leq 0$ to the entire sample space of $\mathbf{X}$ and where $\mathrm{I}[g(\mathbf{x}) \leq 0]$ is an indicator function equal to 1 if $g(\mathbf{x}) \leq 0$ and otherwise equal to zero. Equation 6.20 is in this way seen to yield the expected value of the indicator function $I[g(\mathbf{x}) \leq 0]$. Therefore, if now N realizations of the vector $\mathbf{X}$, i.e. $\mathbf{x}_j$, $j = 1, 2 \ldots, N$ are sampled it follows from sample statistics that:

$$P_F = \frac{1}{N}\sum_{j=1}^{N} I\left[g(\mathbf{x}) \leq 0\right] \tag{6.21}$$

is an unbiased estimator of the failure probability P_F.

The principle of the crude *Monte Carlo simulation* technique rests directly on the application of Eq. 6.21. A large number of realizations of the basic random variables X, i.e. $\mathbf{x}_j$, $j = 1, 2 \ldots, N$ are generated (or simulated) and for each of the outcomes $\mathbf{x}_j$ it is checked whether or not the limit state function taken in $\mathbf{x}_j$ is positive. All the simulations for which this is not the case are counted (n_F) and after N simulations the failure probability P_F may be estimated through:

$$P_F = \frac{n_F}{N} \tag{6.22}$$

which then may be considered a sample expected value of the probability of failure. In fact, for $N \to \infty$ the estimate of the failure probability becomes exact. However, simulations are often costly in computation time and the uncertainty of the estimate is thus of interest. It is easily realized that the coefficient of variation of the estimate is proportional to $1/\sqrt{n_F}$.

6.6.1 Example 6.4: Monte Carlo Simulation

Consider the limit state function $g(\mathbf{x}) \leq 0 = B - K$ with K being the expected costs of a construction project, and B the budget. K is Lognormal distributed with mean value $\mu_K = 50000$ [CHF]. Due to the uncertainties in the performance of the project, the coefficient of variation of the costs is 0.2. The budget is modeled as a Normal distributed random variable with mean value $\mu_B = 100000$ [CHF] and standard deviation $\sigma_B = 5000$ [CHF].

Performing a Monte Carlo simulation with $n = 100000$, 19 simulations are found to lie outside of the positive range; in Fig. 6.4 these simulations lie under the straight line, which represents the linear limit state function. The probability of failure is calculated as $P_F = \frac{n_F}{N} = \frac{19}{100000} = 1.9 \cdot 10^{-4}$. This value will vary due to the rather small number of simulations (10^5) in every new Monte Carlo simulation performed.

If Monte Carlo simulation is pursued to estimate a probability in the order of 10^{-6} it must be expected that approximately 10^8 simulations are necessary to achieve an estimate with a coefficient of variance in the order of 10%. A large number of simulations are thus required using crude Monte Carlo simulation and all refinements of this crude technique have the purpose of reducing the variance of the estimate. Such methods are for this reason often referred to as *variance reduction methods*.

The simulation of the N outcomes of the joint density function in Eq. 6.21 is in principle simple and may be seen as consisting of two steps. Here the steps will be illustrated assuming that the n components of the random vector X are independent.

In the first step, a "*pseudo random*" number with a uniform distribution between 0 and 1 is generated for each of the components in $\mathbf{x}_j$ i.e. x_{ji}, $i = 1, 2, 3, \ldots, n$. The

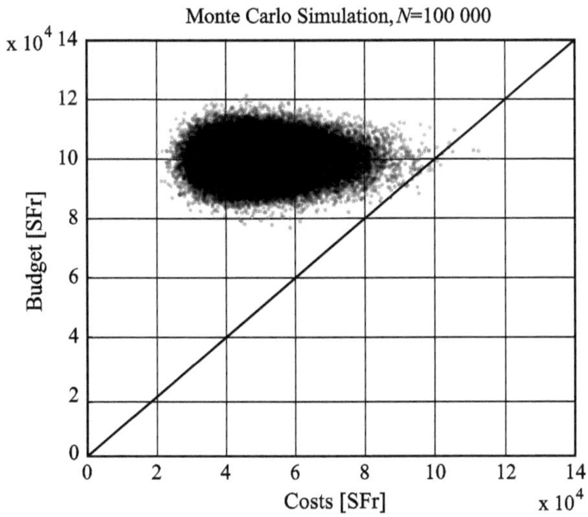

Fig. 6.4 Monte Carlo Simulation with 100000 realizations of the two random variables, budget B and costs K, and evaluation of the limit state function $g(x) \leq 0 = B - K$

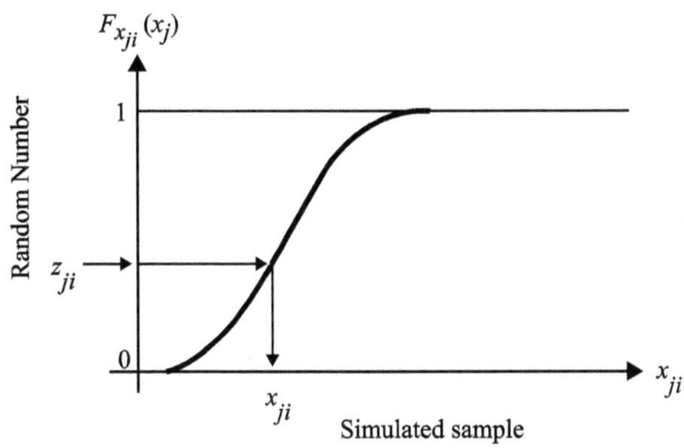

Fig. 6.5 Principle for simulation of a random variable

generation of such numbers may be facilitated by built-in functions of basically all programming languages and spreadsheet software.

In the second step, the outcomes of the "pseudo random" numbers z_{ji} are transformed to outcomes of x_{ji} by:

$$x_{ji} = F_{X_i}^{-1}(z_{ji}) \tag{6.23}$$

where $F_{X_i}(\cdot)$ is the cumulative distribution function for the random variable X_i.

The principle is also illustrated in Fig. 6.5.

This process is continued until all components of the vector $\mathbf{x}_j$ have been generated.

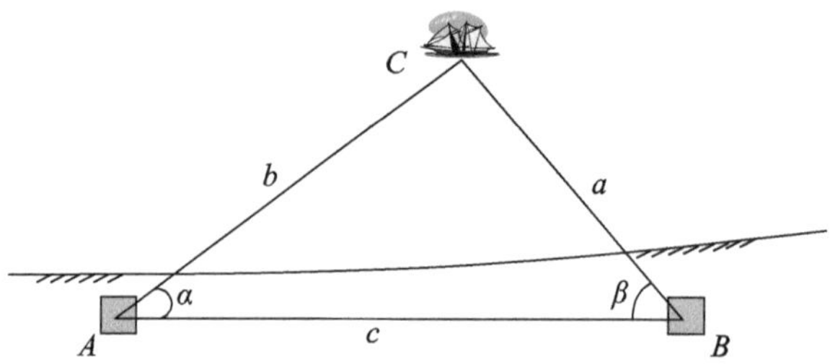

Fig. 6.6 Position determination of a ship

6.7 Self Assessment Questions/Exercises

1. How may failure events be represented in terms of basic random variables in the context of structural reliability theory?
2. What is the geometrical interpretation of the reliability index and how does it relate to the failure probability?
3. Using the Monte Carlo Simulation method, a sample expected value of the probability of failure is estimated. How may the accuracy in the estimation of the probability of failure be increased?
4. Consider a timber beam subjected to an annual maximum bending moment L. The bending strength of the beam R is modeled by a Normal distributed random variable with mean $\mu_R = 30$ kN m and standard deviation $\sigma_R = 5$ kN m and the annual maximum bending moment is modeled by a Normal distributed random variable with mean $\mu_L = 9$ kN m and standard deviation $\sigma_L = 2$ kN m. It is assumed that R and L are independent. The timber beam fails when the applied moment exceeds the bending strength. Calculate the reliability index β and the probability of failure of the timber beam.
5. Consider a steel rod that carries a deterministic load, $S = 35$ kN m. The resistance, R, of the rod is given by the following product: $R = A \cdot f_y$, where A is the area of the rod, equal to 100 mm^2 and f_y is the yield stress modeled as a Normal distributed random variable with mean $\mu_{f_y} = 425 \cdot 10^{-3}$ kN/mm^2 and standard deviation $\sigma_{f_y} = 25 \cdot 10^{-3}$ kN/mm^2. Formulate a safety margin, M, for the steel rod and estimate the reliability of the rod. Draw the probability density function of the safety margin and indicate the safe and failure regions.
6. The position of a ship is measured by two fixed points A and B located at the coast, see Fig. 6.6.

   Angles α and β have been measured from the basis line AB at the same time. Determine the error in b if the following information is provided:

$$\alpha = 6 \text{ km} \pm 0.005 \text{ km}$$

$$\beta = 0.813 \text{ rad} \pm 0.011 \text{ rad}$$

$$\gamma = 1.225 \text{ rad} \pm 0.011 \text{ rad}$$

   where, for instance, $c = 6 \text{ km} \pm 0.005$ km means that the mean value of c is 6 km and the standard deviation of c is 0.005 km.

Chapter 7
Bayesian Decision Analysis

Lecture 13 (Aim of the Present Lecture) The aim of the present lecture is to illustrate how the basic knowledge acquired through the present course provides a strong basis for engineering decision making. By establishing probabilistic engineering models that are consistent with the available knowledge, it is shown how risk or expected consequences can be utilized to identify and rank different engineering decision alternatives. To this end on the basis of a simple example, the three principally different types of decision analysis are introduced, namely the prior-, posterior- and the pre-posterior decision analysis. Whereas the prior and the posterior decision analyses only differ in the available information at hand at the time of decision making and may serve as a direct basis for the planning of engineering activities involving changes of the state of nature, the pre-posterior analysis forms a strong basis for the planning of collection of information through e.g. experiments in the laboratory or in the field.

On the basis of the lecture, it is expected that the reader should acquire knowledge and skills with regard to:

- What must be identified before a decision analysis can be performed?
- What is a utility function and what role does it play in decision making?
- How are risk and utility related?
- How is a decision event tree constructed?
- How can expected utility be calculated based on branching probabilities and consequences?
- How can the uncertainty associated with information be accounted for in decision analysis?
- What is the difference between prior and posterior decision analysis?
- What is the idea behind the pre-posterior decision analysis?
- How can the value of information be assessed?

M.H. Faber, *Statistics and Probability Theory*,
Topics in Safety, Risk, Reliability and Quality 18,
DOI 10.1007/978-94-007-4056-3_7,

7.1 Introduction

The ultimate task for an engineer is to establish a consistent decision basis for the planning, design, manufacturing, construction, operation and management of engineering facilities such that the overall life cycle benefit of the facilities is maximized and such that the given requirements to the safety of personnel and environment specified by legislation or society are fulfilled.

As the available information (regarding, e.g., soil properties, loading, material properties, future operational conditions and deterioration processes in general) is incomplete or uncertain, the decision problem is subject to uncertain information.

The present chapter introduces some fundamental issues of decision making subject to uncertain information. General aspects of decision theory are considered and illustrated using a simple example. Finally, the risk analysis decision problem is defined in general terms within the context of decision theory.

7.2 The Decision/Event Tree

In practical decision problems such as feasibility studies, reassessment of existing structures or decommissioning of facilities that have become obsolete, the number of alternative actions can be extremely large and a framework for the systematic analysis of the corresponding consequences is therefore expedient.

A decision/event tree as illustrated in Fig. 7.1 may conveniently represent the decision problems.

For the purpose of illustration, the decision/event tree in Fig. 7.1 considers the following very simple decision problem. In the specifications for the construction of a production facility using large amounts of fresh water, it is specified that a water source capable of producing at least 100 units of water per day must be available. As it is known that the underground at the location of the planned production facility actually contains a water reservoir, one option is to develop a well locally at the site of the production facility. However, as the capacity of the local water reservoir is not known with certainty another option is to get the water from an alternative location where a suitable well already exists.

The different options are associated with different costs and different potential consequences. The costs of establishing a well locally is assumed to be equal to 10 monetary units. If the already existing well is used, it is necessary to construct a pipeline. As the existing well is located far away from the planned production facility, the associated costs are assumed to be equal to 100 monetary units.

Based on experience from similar geological conditions, it is judged that the probability that a local well will be able to produce the required amount of water is 0.4. Correspondingly, the probability that the well will not be able to fulfill the given requirements is 0.6.

The consequence of establishing locally a well, which turns out to be unable to produce the required amount of water, is that a pipeline to the existing—but

Fig. 7.1 Decision/event tree

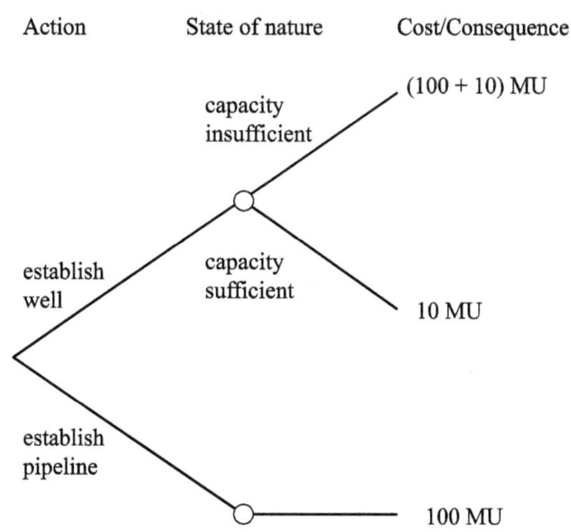

distant—well must be constructed. It is assumed that in this case all the water for the production facility is supplied from this well.

The task is now to analyze the decision problem making consistent use of all the information available to the engineer, including his/her degree of belief in the possible states, his/her subsequent observed data and his/her preferences among the various possible action/state pairs.

To this end, use will be made of the fact that decisions shall be based on expected values of the corresponding consequences. This issue is addressed further in the following section.

7.3 Decisions Based on Expected Values

Consider the simple case where the engineer must choose between actions a_1 and a_2. The consequence of action a_2 is C with certainty whereas the consequence of action a_1 is uncertain. The state of nature may be θ_1, in which case the consequence is A and the state of nature may be θ_2 in which case the consequence is B. The decision/event tree is illustrated in Fig. 7.2.

Before the true state of nature is known, the optimal decision depends upon the likelihood of the various states of the nature θ and the consequences A, B and C.

A further analysis of the decision problem requires the numerical assessment of the preferences of the decision maker. It is assumed that the decision maker prefers B to A, C to A, and B to C. This statement of preferences may be expressed by any function u such that:

$$u(B) > u(C) > u(A) \tag{7.1}$$

The task is to find a particular function u, namely the utility function, such that it is logically consistent to decide between a_1 and a_2 by comparing $u(C)$ with the

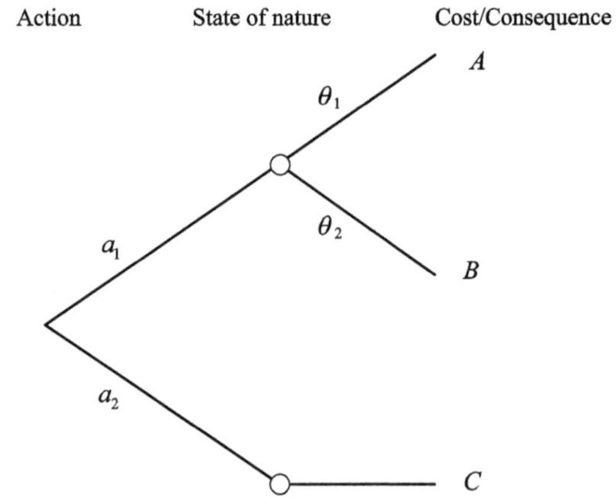

Fig. 7.2 Decision/event tree illustrating a basic decision problem

expected value of the utility of the action a_1, namely:

$$pu(A) + (1 - p)u(B) \tag{7.2}$$

where p is the probability that the state of nature is θ_1.

Assuming that $u(A)$ and $u(B)$ have been given appropriate values, the question is—what value should $u(C)$ have in order to make the expected value a valid decision criterion? If the probability of θ_1 being the state of nature p is equal to 0, the decision maker would choose a_1 over a_2 because she/he prefers B to C. On the other hand, if the probability of θ_1 being the state of nature is equal to 1 she/he would choose a_2 over a_1. For a value of p somewhere between 0 and 1, the decision maker will be indifferent to choosing a_1 over a_2. This value p^* may be determined and $u(C)$ is assigned as:

$$u(C) = p^*u(A) + (1 - p^*)u(B) \tag{7.3}$$

From Eq. 7.3 it is seen that $u(C)$ will lie between $u(A)$ and $u(B)$ for all choices of p^* and therefore the utility function is consistent with the stated preferences. Furthermore, it is seen that the decision maker should choose the action a_1 to a_2 only if the expected utility given this action $E[u|a_1]$ is greater than $E[u|a_2]$. This is realized by noting that for all p greater than p^* and with $u(C)$ given by Eq. 7.3. There is:

$$\begin{gathered} u(C) > pu(A) + (1 - p)u(B) \\ \Updownarrow \\ p^*u(A) + (1 - p^*)u(B) > pu(A) + (1 - p)u(B) \\ \Updownarrow \\ u(B) + (u(A) - u(B))p^* > u(B) + (u(A) - u(B))p \end{gathered} \tag{7.4}$$

This means that if $u(C)$ is properly assigned in consistency with the decision makers stated preferences (i.e. B preferred to C preferred to A) and the indifference probability p^*, the ranking of the expected values of the utility determines the ranking of actions.

7.4 Decision Making Subject to Uncertainty

Having formulated the decision problem in terms of a decision/event tree, with proper assignment of utility and probability structure, the numerical evaluation of decision alternatives may be performed.

Depending on the state of information at the time of the decision analysis, three different analysis types are distinguished, namely prior analysis, posterior analysis and pre-posterior analysis. Each of these are important in practical applications of decision analysis and are therefore discussed briefly in the following.

7.5 Decision Analysis with Given Information—Prior Analysis

When the utility function has been defined and the probabilities of the various state of nature corresponding to different consequences have been estimated, the analysis is reduced to the calculation of the expected utilities corresponding to the different action alternatives. In the following examples the utility is represented in a simplified manner through the costs whereby the optimal decisions now should be identified as the decisions minimizing expected costs, which then is equivalent to maximizing expected utility.

At this stage, the probabilistic description $P[\theta]$ of the state of nature θ is usually called a prior description and called $P'[\theta]$.

To illustrate the *prior decision analysis* the decision problem from Sect. 7.2 is considered again. The decision problem is stated as follows. The decision maker has a choice between two actions:

a_1: Establish a new well
a_2: Establish a pipeline from an existing well

The possible states of nature are the following two:

θ_1: Capacity insufficient
θ_2: Capacity sufficient

The prior probabilities are:

$P'[\theta_1] = 0.60$
$P'[\theta_2] = 0.40$

Based on the prior information alone, it is easily seen that the expected cost $E'[C]$ amounts to:

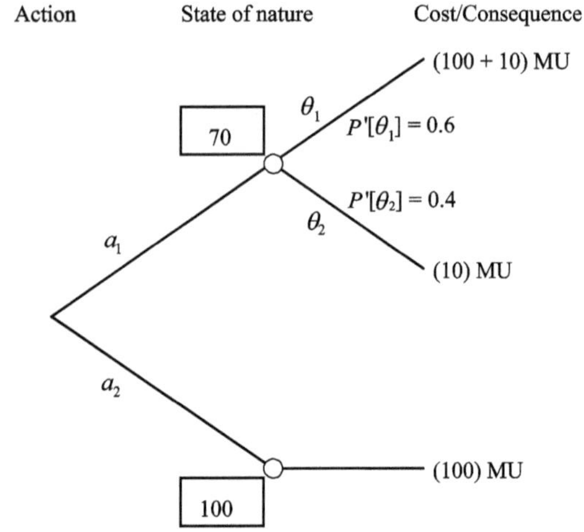

Fig. 7.3 Simple decision problem with assigned prior probabilities and utility

$$\begin{aligned} E'[C] &= \min\left\{P'[\theta_1]\cdot(100+10)+P'[\theta_2]\cdot 10;\ 100\right\} \\ &= \min\{70;\ 100\} = 70 \text{ monetary units} \end{aligned}$$

The decision/event tree is illustrated in Fig. 7.3 together with the expected costs (in boxes).

It is seen that action alternative a_1 yields the smallest expense (largest expected utility) so this action alternative is the optimal decision.

7.6 Decision Analysis with Additional Information—Posterior Analysis

When additional information becomes available, the probability structure in the decision problem may be updated. Having updated the probability structure the decision analysis is unchanged in comparison to the situation with given—prior information.

Given the result of an experiment z_k, the updated probability structure (or just the posterior probability) is called $P''[\theta]$ and may be evaluated by use of the Bayes' rule:

$$P''[\theta_i] = \frac{P[z_k|\theta_i]P'[\theta_i]}{\sum_j P[z_k|\theta_j]P'[\theta_j]} \tag{7.5}$$

which may be explained as:

$$\begin{aligned} &\begin{pmatrix}\text{Posterior probability of } \theta_i \\ \text{with given sample outcome}\end{pmatrix} \\ &= \begin{pmatrix}\text{Normalizing} \\ \text{constant}\end{pmatrix}\cdot\begin{pmatrix}\text{Sample likelihood} \\ \text{given } \theta_i\end{pmatrix}\cdot\begin{pmatrix}\text{prior probability} \\ \text{of } \theta_i\end{pmatrix} \end{aligned} \tag{7.6}$$

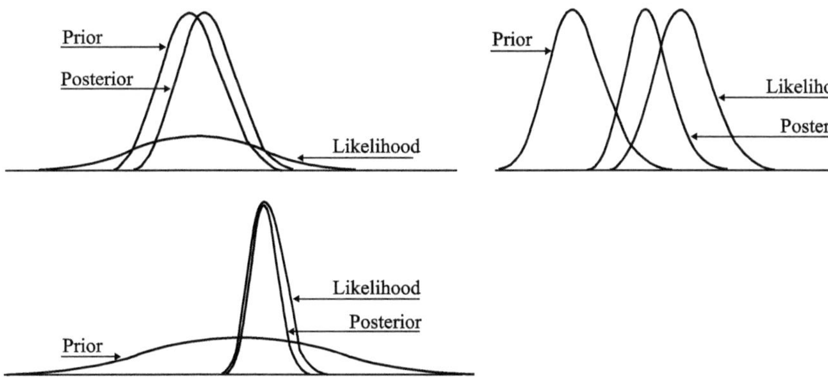

Fig. 7.4 Illustration of updating of probability structures

The normalizing factor is to ensure that $P''[\theta_i]$ forms a proper probability. The mixing of new and old information appears through the sample likelihood $P[z_k|\theta_i]$ and the prior probability $P'[\theta_i]$. The likelihood is the probability of obtaining the observation z_k given the true state of nature θ_i.

In Fig. 7.4 an illustration is given of corresponding prior and posterior probability density functions together with likelihood functions. In the first case, the prior information is strong and the likelihood is weak (e.g. due to a small sample size). In the second case, the prior information and the likelihood are of comparable strength. In the last case, the prior information is relatively weak in comparison to the likelihood.

To illustrate the *posterior decision analysis*, the water supply decision problem is considered again.

It is assumed that information about the capacity of the local reservoir can be estimated by the implementation of a less expensive test well and a subsequent pump test. It is assumed that the cost of establishing a test well is equal to 1 monetary unit. However, the information obtained from the pump test is only indicative as the result of the difference in scale from the test well to the planned local well.

It is further assumed that the pump test can provide the following different information—i.e. indicators regarding the capacity of the local reservoir.

The capacity of the reservoir is:

- Larger than the given production requirements by 5% i.e. larger than 105 water volume units per day.
- Less than 95% of the required water production, i.e. less than 95 water volume units.
- Between 95 and 105 water units.

The information from the pump test is subject to uncertainty and the likelihood of the actual capacity of the local reservoir given the three different indications described above are given in Table 7.1.

Given that a test well is established and a trial pump test conducted with the result that a capacity smaller than 95 water volume units is indicated, a posterior

Table 7.1 Likelihood of the true capacity of the reservoir given the trial pump test results

Indicators	True capacity of the reservoir	
	θ_1: less than 100	θ_2: more than 100
I_1 : capacity > 105	0.1	0.8
I_2 : capacity > 95	0.7	0.1
$I_3 : 95 \leq$ capacity ≤ 105	0.2	0.1

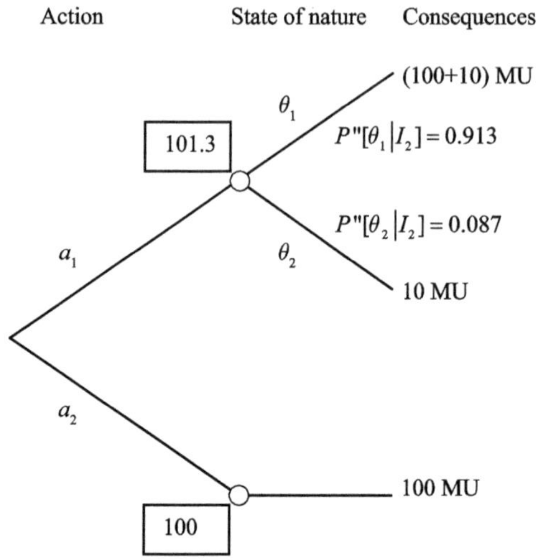

Fig. 7.5 Illustration of decision/event tree for water supply decision problem

decision analysis can be performed to identify whether the optimal decision is to establish a well locally or if it is more optimal to construct a pipeline to the existing well.

Therefore, the posterior probabilities given the new information $P''[\theta|z]$ can be given as:

$$P''[\theta_1|I_2] = \frac{P[I_2|\theta_1]P'[\theta_1]}{P[I_2|\theta_1]P'[\theta_1] + P[I_2|\theta_2]P'[\theta_2]}$$
$$= \frac{0.7 \cdot 0.6}{0.7 \cdot 0.6 + 0.1 \cdot 0.4} = \frac{0.42}{0.46} = 0.913$$

$$P''[\theta_2|I_2] = \frac{P[I_2|\theta_2]P'[\theta_2]}{P[I_2|\theta_1]P'[\theta_1] + P[I_2|\theta_2]P'[\theta_2]}$$
$$= \frac{0.1 \cdot 0.4}{0.7 \cdot 0.6 + 0.1 \cdot 0.4} = \frac{0.04}{0.46} = 0.087$$

which are also shown in Fig. 7.5. Having determined the updated probabilities the posterior expected values $E''[C|I_2]$ of the utility corresponding to the optimal action alternative is readily obtained as:

$$\begin{aligned} E''[C|\,I_2] &= \min\left\{P''[\theta_1|\,I_2]\cdot(100+10)+P''[\theta_2|\,I_2]\cdot 10;\,100\right\} \\ &= \min\{101.3;\,100\} = 100 \text{ monetary units} \end{aligned}$$

and indicated in boxes in Fig. 7.5.

Considering the additional information, it is seen that the optimal decision has been switched to action a_2 as compared to the prior decision analysis.

7.7 Decision Analysis with 'Unknown' Information—Pre-posterior Analysis

Often the decision maker has the possibility to 'buy' additional information through an experiment before actually making his/her choice of action. If the cost of this information is small in comparison to the potential value of the information, the decision maker should perform the experiment. If several different types of experiments are possible, the decision maker must choose the experiment yielding the overall largest expected value of utility.

If the example from the previous sections is considered again, the decision problem could be formulated as a decision to decide whether or not to perform the trial pump tests. The *pre-posterior decision analysis* facilitates this.

The situation prior to performing the experiment has already been considered in Sect. 7.5. There it was found that the expected cost based entirely on the prior information $E'[C]$ is 70 monetary units.

In this situation the experiment is planned but the result is still unknown. In this situation the expected cost, disregarding the experiment cost, can be found as:

$$E[C] = \sum_{i=1}^{n} P'[I_i]E''[C|\,I_i] = \sum_{i=1}^{n} P'[I_i] \min_{j=1,\ldots,m}\left\{E''[C(a_j)|\,I_i]\right\} \tag{7.7}$$

where n is the number of different possible experiment findings and m is the number of different decision alternatives. In Eq. 7.7 the only new term in comparison to the previous section is $P'[I_i]$ which may be calculated by:

$$P'[I_i] = P[I_i|\,\theta_1]P'[\theta_1] + P[I_i|\,\theta_2]P'[\theta_2] \tag{7.8}$$

With reference to Sects. 7.5 and 7.6 the prior probabilities of obtaining the different indications by the tests are $P'[I_1]$, $P'[I_2]$ and $P'[I_3]$ and given by:

$$\begin{aligned} P'[I_1] &= P[I_1|\,\theta_1]P'[\theta_1] + P[I_1|\,\theta_2]P'[\theta_2] = 0.1\cdot 0.6 + 0.8\cdot 0.4 = 0.38 \\ P'[I_2] &= P[I_2|\,\theta_1]P'[\theta_1] + P[I_2|\,\theta_2]P'[\theta_2] = 0.7\cdot 0.6 + 0.1\cdot 0.4 = 0.46 \\ P'[I_3] &= P[I_3|\,\theta_1]P'[\theta_1] + P[I_3|\,\theta_2]P'[\theta_2] = 0.2\cdot 0.6 + 0.1\cdot 0.4 = 0.16 \end{aligned}$$

The posterior expected cost terms in Eq. 7.7 are found to be:

$$E''[C|\,I_1] = \min\left\{P''[\theta_1|\,I_1]\cdot(100+10) + P''[\theta_2|\,I_1]\cdot 10;\,100\right\}$$

$$
\begin{aligned}
&= \min\{0.158 \cdot 110 + 0.842 \cdot 10;\ 100\} \\
&= \min\{25.8;\ 100\} = 25.8 \text{ monetary units} \\
E''[C|\,I_2] &= \min\left\{P''[\theta_1|\,I_2] \cdot (100+10) + P''[\theta_2|\,I_2] \cdot 10;\ 100\right\} \\
&= \min\{0.913 \cdot 110 + 0.087 \cdot 10;\ 100\} \\
&= \min\{101.3;\ 100\} = 100 \text{ monetary units} \\
E''[C|\,I_3] &= \min\left\{P''[\theta_1|\,I_3] \cdot (100+10) + P''[\theta_2|\,I_3] \cdot 10;\ 100\right\} \\
&= \min\{0.75 \cdot (100+10) + 0.25 \cdot 10;\ 100\} \\
&= \min\{85;\ 100\} = 100 \text{ monetary units}
\end{aligned}
$$

where the posterior probabilities $P''[\theta_i\,|\,I_1]$ and $P''[\theta_i\,|\,I_2]$ are determined as already shown in Sect. 7.6 for $P''[\theta_i\,|\,I_3]$. The expected cost corresponding to the situation where the experiment with the experiment costs C_P is therefore:

$$
\begin{aligned}
E[C] &= E''[C|\,I_1]\,P'[I_1] + E''[C|\,I_2]\,P'[I_2] + E''[C|\,I_3]\,P'[I_3] \\
&= (25.8 + C_P) \cdot 0.38 + (100 + C_P) \cdot 0.46 + (85 + C_P) \cdot 0.16 \\
&= (69.4 + C_P) \text{ monetary units}
\end{aligned}
$$

By comparing this result with the expected cost corresponding to the prior information, it is seen that the experiment should be performed if the cost of the experiment is less than 0.6 monetary units:

$$
E'[C] - E[C] = 70 - (69.4 + C_P) = 0.6 - C_P
$$

The 0.6 monetary units in this case correspond to the value of information.

7.8 The Risk Treatment Decision Problem

Having introduced the fundamental concepts of decision theory, it will now be considered how these carry over to the principally different types of risk analysis.

The simplest form of risk analysis, i.e. a simple evaluation of the risks associated with a given activity and/or decision alternative may be related directly to the prior decision analysis. In the prior analysis the risk is evaluated on the basis of statistical information and probabilistic modeling available prior to any decision and/or activity. A simple decision/event tree in Fig. 7.6 illustrates the prior analysis. In a prior analysis the risk for each possible activity/option may e.g. be evaluated as:

$$
R = E[U] = \sum_{i=1}^{n} P_i\,C_i \tag{7.9}
$$

where R is the risk, U the utility, P_i is the ith branching probability and C_i the consequence of the event of branch i.

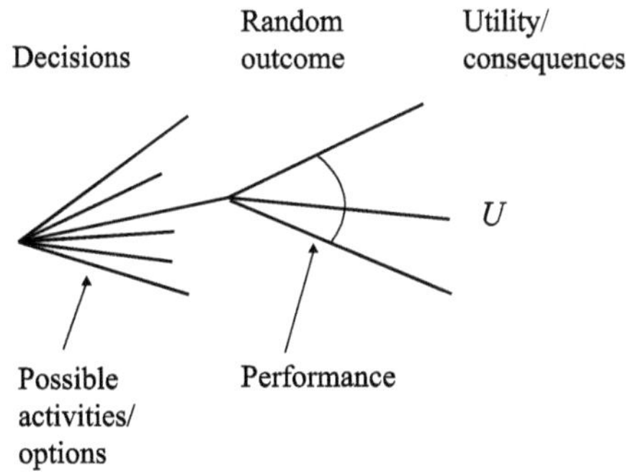

Fig. 7.6 Decision/event tree for prior and posterior decision analysis

A prior analysis, in fact, corresponds closely to the assessment of the risk associated with a known activity. A *prior analysis thus forms the basis for the comparison of risks between different activities.*

A posterior analysis is, in principle, of the same form as the prior analysis; however, changes in the branching probabilities and/or the consequences in the decision/event tree reflect that the considered problem has been changed as an effect of risk reducing measures, risk mitigating measures and/or collection of additional information.

A posterior analysis may thus be used to evaluate the effect of activities, which factually have been performed. For example, for assessment of existing facilities, the testing and inspection of the "as built" facility would be expected to reveal many gross design and construction errors, leading to a more accurate reliability analysis.

A pre-posterior analysis may be illustrated by the decision/event tree shown in Fig. 7.7. *Using pre-posterior analysis, optimal decisions with regard to activities that may be performed in the future, e.g. the planning of risk reducing activities and/or collection of information may be identified.* An important prerequisite for pre-posterior analysis is that decision rules need to be formulated for specifying the future actions that will be taken on the basis of the results of the planned activities.

In a pre-posterior analysis, the optimal investigation a^* is identified through:

$$\min_a E'_Z\left[E''_Z[C(a(z),z)]\right] = \min_a E'_Z\left[\sum_{i=1}^{n} P_i{}''(a(z),z)C_i(a(z))\right] \tag{7.10}$$

where $a(z)$ are the different possible actions that can be taken on the basis of the result of the considered investigation z and $E[\cdot]$ is the expected value operator. $'$ and $''$ refer to the probabilistic description of the events of interest based on prior and posterior information respectively. In Eq. 7.10 the expected utility has been associated only with expected costs; hence the optimal decision is identified through a minimization. If utility, more generally, is associated with expected benefits, the optimization should be performed through maximization.

Pre-posterior analyses form a strong decision support tool and have been intensively used for the purpose of risk based inspection planning. However, so far pre-posterior decision analysis has been grossly overlooked in risk assessments.

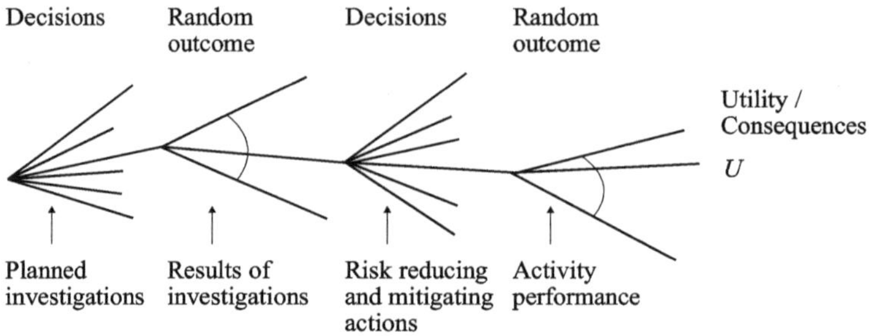

Fig. 7.7 Decision/event tree for pre-posterior decision analysis

It is important to note that the probabilities for the different events represented in the prior or posterior decision analysis may be assessed by logic tree analysis, classical reliability analysis and structural reliability analysis or any combination of these. The risk analysis thus in effect includes all these aspects of systems and component modeling in addition to providing the framework for the decision making.

7.9 Self Assessment Questions/Exercises

1. What must be identified before a decision analysis can be performed?
2. What is a utility function?
3. What is the difference between prior and posterior decision analysis?
4. What is the idea behind the pre-posterior decision analysis?
5. After the occurrence of an event of heavy snowfall, you need to decide whether to clean up the snow from a roof or not. In the following some information is provided to assist you in the decision making process.
   The clean up of the roof can be made by the local fire department. This option is associated with a cost equal to 4000 SFr. In the case of collapse of the roof due to the snow load, the associated cost is equal to 1000000 SFr.
   The probability of collapse of the roof has been estimated using First Order Reliability Methods (FORM). If the snow is dry (denoted asSD), the probability of collapse is: $P_f(SD) = 10^{-3}$. If the snow is wet (denoted as SW), the probability of collapse is: $P_f(SW) = 6.2 \cdot 10^{-3}$. Based on experience, it can be said that tells that the probability of having wet snow on the roof is $P(SW) = 0.6$. In case where there is no snow (denoted as SN), on the roof the probability of collapse is equal to: $P_f(SN) = 5 \cdot 10^{-4}$.
   With the aid of a melting test, the characteristics of the snow can be ascertained. The test costs 1000 SFr. The test indication of the snow property, I_{SD} or I_{SW}, is correct only in 75% of the cases. As the uncertainty of the indication is independent of the snow properties, $P(I_{SD}|\,SD) = P(I_{SW}|\,SW)$.
   Build an appropriate event tree and use it to find out which decision alternative—"clean up", "do nothing" or "carry out test" is the most beneficial one in terms of cost.

Appendix A
Answers to Self Assessment Questions

A.1 Chapter 1

1.1 According to the so-called Brundtland Commission [5], sustainable development is defined as development "that meets the needs of the present without compromising the ability of future generations to meet their own needs". Consideration of sustainable development leads to sustainable decision making which may be understood as based on a joint consideration of society, economy and environment (*for more see Sect.* 1.1).

1.2 A beneficial engineered facility is understood as: being economically efficient in serving a specific purpose, fulfilling given requirements with regard to the safety of the personnel, and fulfilling given requirements to limit the adverse effects of the facility on the environment (*for more, see Sect.* 1.2).

1.3 As discussed in Sect. 1.3 when considering an activity with only one event with potential consequences C, the risk R is the probability P that this event will occur multiplied with the consequences given the event occurs i.e.:

$$R = PC$$

1.4 The term "acceptable risks" points out to "what is one prepared to invest and/or pay for the purpose of getting a potential benefit" (*for more, see Sect.* 1.2).

1.5 As discussed in Sect. 1.3, the risk of an event is calculated by Eq. 1.1 such as: $R = PC$. Hence the given table can easily be completed and it can be seen that event 3 is associated with the higher risk.

Event	1	2	3
Event probability	0.1	0.01	0.2
Consequences	100 SFr	500 SFr	100 SFr
Risk	10	5	20

M.H. Faber, *Statistics and Probability Theory*,
Topics in Safety, Risk, Reliability and Quality 18,
DOI 10.1007/978-94-007-4056-3,

A.2 Chapter 2

2.1 The estimation is based on the frequentistic interpretation of probability. In the frequentistic interpretation, the probability $P(A)$ is simply the relative frequency of occurrence of the event A as observed in an experiment with n trials. It is mathematically defined as (*for more, see Sect.* 2.2):

$$P(A) = \lim \frac{N_A}{n_{\text{exp}}} \quad \text{for } n_{\text{exp}} \to \infty$$

2.2 Following the Bayes' rule (*see Sect.* 2.5), the conditional probability of the event E_1 given that the event E_2 has occurred is written as:

$$P(E_1|E_2) = \frac{P(E_1 \cap E_2)}{P(E_2)}$$

2.3 In probability theory, the probability $P(A)$ of an event A can take any value within the following boundaries:
$0 \leq P(A) \leq 1$ ✓
$-1 \leq P(A) \leq 1$ ☐
$-\infty \leq P(A) \leq \infty$ ☐

2.4 If the intersection of two events, A and B corresponds to the empty set $\emptyset$, i.e. $A \cap B = \emptyset$, the two events are:
Mutually exclusive. ✓
Independent. ☐
Empty events. ☐

2.5 Which of the following expressions is(are) correct?
The probability of the union of two events A and B is equal to the sum of the probability of event A and the probability of event B, given that the two events are mutually exclusive. ✓
The probability of the union of two events A and B is equal to the probability of the sum of event A and event B, given that the two events are mutually exclusive. ☐
The probability of the intersection of two events A and B is equal to the product of the probability of event A and the probability of event B, given that the two events are mutually exclusive. ☐
The probability of the intersection of two events A and B is equal to the product of the probability of event A and the probability of event B, given that the two events are independent. ✓

2.6 The probability of the intersection of two mutually exclusive events is equal to:
The product of the probabilities of the individual events. ☐
The sum of the probabilities of the individual events. ☐
The difference between the probabilities of the individual events. ☐
One (1). ☐
Zero (0). ✓

2.7 Within the theory of sample spaces and events, which of the following statements is(are) correct?
An event A is defined as a subset of a sample space Ω. ✓
A sample space Ω is defined as a subset of an event A. ☐

2.8 The probability of the union of two not mutually exclusive events A and B is given as: $P(A \cup B) = P(A) + P(B) - P(A \cap B)$. It is provided that the probability of event A is equal to 0.1, the probability of event B is 0.1 and the probability of event B given event A, i.e. $P(B|A)$ is 0.8. Which result is correct?
$P(A \cup B) = -0.6$ ☐
$P(A \cup B) = 0.12$ ✓
$P(A \cup B) = 0.04$ ☐

2.9 For an event A in the sample space Ω, event $\bar{A}$ represents the complementary event of event A. Which of the following hold?
$A \cup \bar{A} = \Omega$ ✓
$A \cap \bar{A} = \Omega$ ☐
$A \cup \bar{A} = \emptyset$ ☐

2.10 The commutative, associative and distributive laws describe how to:
Operate with intersections of sets. ✓
Operate with unions of sets. ✓
None of the above. ☐

2.11 Following the principles explained in Sect. 2.5 we obtain:
a) The table is completed as follows:

SNF final decision D_i	Dr. Beispiel's indicative assessment I_j		
	I_1	I_2	I_3
D_1	$P(I_1\|D_1) = 0.86$	$P(I_2\|D_1) = 0.1$	$P(I_3\|D_1) = 0.04$
D_2	$P(I_1\|D_2) = 0.2$	$P(I_2\|D_2) = 0.74$	$P(I_3\|D_2) = 0.06$
D_3	$P(I_1\|D_3) = 06$	$P(I_2\|D_3) = 0.1$	$P(I_3\|D_3) = 0.9$

b) Using the Bayes' Rule the probability that the final decision made by SNF is the same with the indicative assessment of Dr. Beispiel is:

$$
\begin{aligned}
&P(D_2 | I = D_2) \\
&= \frac{P(I = D_2|D_2)P(D_2)}{\sum_{i=1}^{3} P(I = D_2|D_i)P(D_i)} \\
&= \frac{P(I = D_2|D_2)P(D_2)}{P(I = D_2|D_1)P(D_1) + P(I = D_2|D_2)P(D_2) + P(I = D_2|D_3)P(D_3)} \\
&= \frac{0.74 \cdot 0.35}{(0.1 \cdot 0.45) + (0.74 \cdot 0.35) + (0.1 \cdot 0.2)} \\
&= 0.799
\end{aligned}
$$

A.3 Chapter 3

3.1 The main purpose of the use of descriptive statistics is to assess the characteristics and the level of uncertainty of a given quantity of interest without assuming anything in terms of the degree or nature of the randomness underlying the data analyzed (*see also Sect.* 3.1).

3.2 By definition, the sample coefficient of correlation may lie in the interval $[-1; 1]$. In both extreme cases, there are linear relationships between two data sets (*see also Sect.* 3.2).

3.3 The interval width plays a role for the resolution of the representation of the observations. If the interval width is too large, the histogram tells little about relative occurrences of individual phenomena. If the width is too small, the relative occurrences in each interval fluctuate due to the random nature of the phenomena (*see also Sect.* 3.3).

3.4 As discussed in Sect. 3.3, five characteristics of a data set are normally presented in a Tukey box plot: the lower adjacent value, the lower quartile, the median, the upper quartile and the upper adjacent value. Outside values can also be shown on a Tukey box plot.

3.5 Q-Q plots provide an efficient means of comparison of observations of two different data sets (*see also Sect.* 3.3).

3.6 Provide an estimate of the correlation coefficient of the data sets plotted in the following figure.

A $r_{XY} \approx +0.8$ B $r_{XY} \approx +0.4$

C $r_{XY} \approx 0$ D $r_{XY} \approx -0.4$

3.7 A number of statistical terms are shown in the following table. Check if the terms have something to do with (a) location parameter, (b) dispersion parameter or (c) none of the above.

	a	b	c
Mean	✓		
Quartile	✓		
Sample size			✓
Median	✓		
Standard deviation		✓	
Coefficient of variation		✓	

3.8 Measurements were taken of the concrete cover depth of a bridge column. The histogram of the measured values has been plotted.

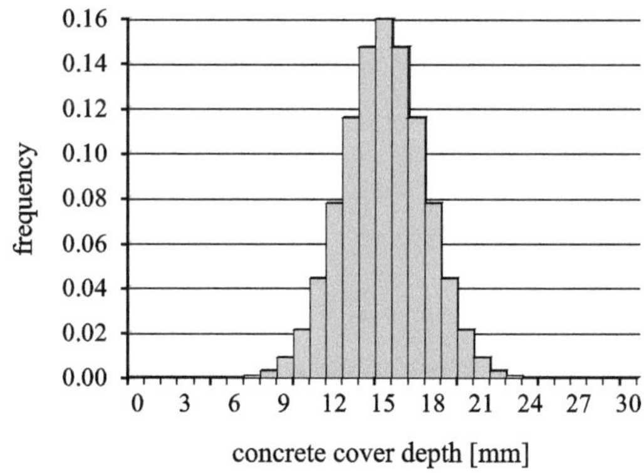

The sample mean, $\bar{x}$, is equal to 0.16 mm. ☐
The sample mean, $\bar{x}$, is equal to 15 mm. ✓
The mode of the data set is equal to 15 mm. ✓

3.9 Which of the following are features of a symmetrical probability density function?
The variance is equal to the coefficient of variation. ☐
The mode is equal to the median. ✓
The skewness is equal to zero. ✓
None of the above. ☐

A.4 Chapter 4

4.1 Inherent natural variability may be interpreted simply as the uncertainty which cannot be reduced by means of collection of additional information. This definition implies that the amount of uncertainty due to inherent natural variability depends on the models applied in the formulation of the engineering problem. Presuming that a refinement of models corresponds to looking in more detail at the problem at hand, one could say that the uncertainty structure influencing a problem is scale dependent. The type of uncertainty associated with the state of knowledge has a time dependency. In principle, if the observation is perfect without any errors the knowledge about the phenomenon is perfect. The modeling of the same phenomenon in the future, however, is uncertain as this involves models subject to natural variability, model uncertainty and statistical uncertainty. The above discussion shows another interesting effect that the uncertainty associated with a model concerning the future transforms from a mixture of aleatory and epistemic uncertainty to a purely epistemic uncertainty when the modeled phenomenon is observed (*see also Sect.* 4.2).

4.2 Epistemic uncertainty involves statistical uncertainty and model uncertainty. Epistemic uncertainty may be reduced by e.g. collecting additional information. On the other hand, aleatory uncertainty is related to the random nature of phenomena, and thus cannot be reduced by collecting information (*see also Sect.* 4.2).

4.3 A continuous random variable is a random variable which can take on any value (*see also Sect.* 4.3).

4.4 With the help of Eqs. 4.16 and 4.18,
   a) $E[a+bX]=a+bE[X]$
   b) $Var[a+bX]=b^2 \cdot Var[X]$

4.5 The required characteristics of the random variable are shown in the following illustrations.

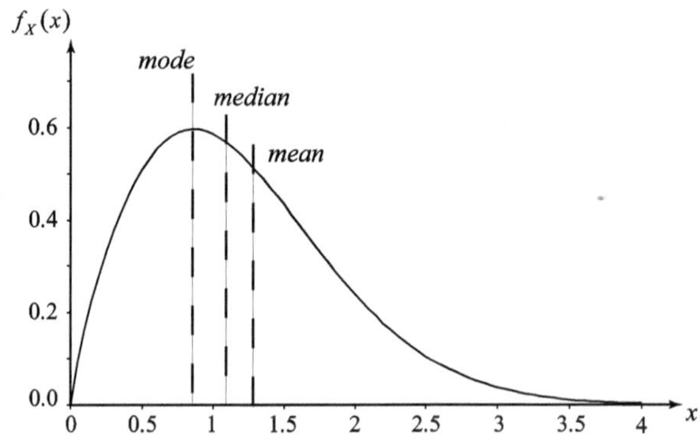

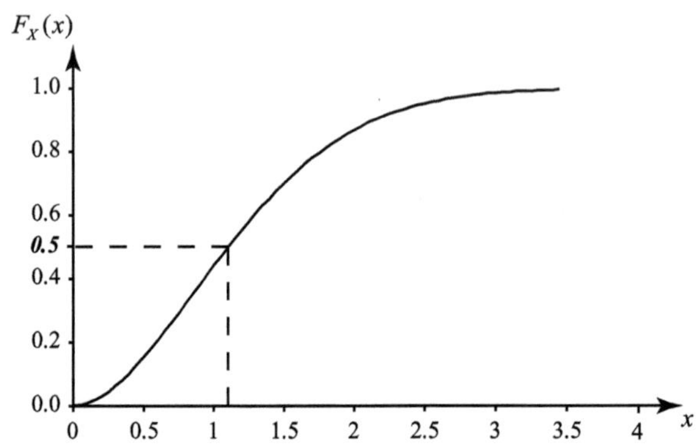

4.6 According to the central limit theorem, "the probability distribution for the sum of a number of random variables approaches the Normal distribution as the number becomes large".

4.7 The standard Normal distribution is a special case of the Normal distribution. A standardized random variable is a random variable that has been transformed such as its expected value is equal to zero and its variance is equal to one (*see also Eq.* 4.48).

4.8 A sequence of experiments with only two possible mutually exclusive outcomes is called a sequence of Bernoulli trials. Typically, the two possible events of a Bernoulli trial are referred to as a success or a failure (*see also Sect.* 4.4).

4.9 Poisson process is a family of discrete processes, which may be used for modeling the number of occurrences of events (*see also Sect.* 4.3).

4.10 The probability of exceeding the value of 5 is calculated as:

$$P(X > 5) = \int_{5}^{10} \frac{3}{1000} x^2\, dx = \left. \frac{x^3}{1000} \right|_{5}^{10} = 0.875$$

4.11 The probability that the engine breaks down within 2 years after placed in operation is calculated as:

$$P(T \leq 2 \text{ years}) = F_T(2) = 1 - e^{(-\frac{2}{10})} = 0.181$$

4.12 The probability of no snowfall in the next year is equal to 0.067. The probability of exactly 5 snowfalls in the next year is equal to 0.176. The probabilities are calculated as:

$$P(X = 0) = \frac{(5 \cdot 1)^0}{0!} e^{-5 \cdot 1} = 0.0067 \quad \text{and} \quad P(X = 5) = \frac{(5 \cdot 1)^5}{5!} e^{-5 \cdot 1} = 0.176$$

A.5 Chapter 5

5.1 The data seem to fit well on a straight line and hence the assumption of a Gumbel distribution can be accepted by the engineer.

5.2 The Maximum Likelihood Method (MLM) enables engineers to calculate the distribution parameters of a random variable on the basis of data. Which of the following statement(s) is(are) correct?
The MLM provides point estimates of the distribution parameters. ✓
The MLM provides information about the uncertainty associated with the estimated parameters. ✓
The MLM provides no information about the uncertainty associated with the estimated parameters. ☐

5.3 From past experience it is known that the shear strength of soil can be described by a Lognormal distribution. 15 samples of soil are taken from a site and an engineer wants to use the data in order to estimate the parameters of the Lognormal distribution. The engineer: may use a probability paper to estimate the parameters of the Lognormal distribution. ✓
may use the maximum likelihood method to estimate the parameters of the Lognormal distribution. ✓
may use the method of moments to estimate the parameters of the Lognormal distribution. ✓
None of the above. ☐

5.4 The procedure of establishing a probabilistic model, as described in Sect. 5.1, consists of five steps:
1) Assessment and statistical quantification of the available data
2) Selection of distribution function
3) Estimation of distribution parameter
4) Model verification and
5) Model updating.

5.5 The probability that the sample average of the steel yield stress will lie within an interval of ± 9.8 MPa of the true mean value μ_X is 0.95 (*see also Sect.* 5.3, *Eq.* 5.22).

5.6 The hypothesis testing procedure, as described also in Sect. 5.4, consists of the following steps/actions:
1) Formulate a null hypothesis, H_0
2) Formulate an operating rule, H_1
3) Select a significance level, α
4) Identify the value resulting in a probability α of performing a Type I error
5) Perform the testing, obtain the sample statistic
6) Judge the null hypothesis.

5.7 An engineer tests the null hypothesis that the mean value of the concrete cover depth of a concrete structure corresponds to design assumptions. In a preliminary assessment a limited number of measurements of the concrete cover depth are made, and after performing the hypothesis test the engineer accepts the null hypothesis. After a few years, a comprehensive survey of the concrete cover depth is carried out, i.e. many measurements are made. The survey shows that the mean value of the concrete cover depth does not fulfill the design assumptions. Which of the following statement(s) is(are) correct?
In the preliminary survey the engineer has performed a Type I error. ☐
In the preliminary survey the engineer has performed a Type II error. ✓
In the preliminary survey the engineer has performed a Type I and a Type II error. ☐

5.8 Probability papers are useful for checking the plausibility of a selected distribution family. A probability paper for a given distribution family is constructed such that the cumulative probability distribution function (or the complement) for that distribution family will have the shape of a straight line when plotted on the paper. A probability paper is thus constructed by a non-linear transformation of the y-axis (*see also Sect.* 5.2).

5.9 It is suggested that the data are lumped in a way that each interval contains about 5 or more observations. If the data are realizations from a continuous distribution function, then they must be descritized (*see also Sect.* 5.9).

5.10 Both tests are used to assess the goodness of fit of the assumed model with data. The Chi-square test is used basically for discrete distribution functions, while the Kolmogorov-Smirnov test is used for continuous distribution functions. However, by the descretization of a continuous distribution function, the Chi-square test can be used also for the continuous distribution functions. Another difference is that whereas the Chi-square test can be applied for the cases where the distribution parameters are already estimated from the same data, the Kolmogorov-Smirnov test cannot be applied when the distribution parameters are estimated from the same data (*see also Sect.* 5.9).

5.11 Following the introduction of the χ^2 goodness of fit test in Sect. 5.9, the number of degrees of freedom of the Chi-square sample statistic ε_m^2 are $k-1=3-1=2$, where k the number of intervals into which the samples were divided. The null hypothesis H_o that X follows a Normal distribution with the given parameters can be tested using the following operating rule:

$$P(\varepsilon_m^2 \geq \Delta) = \alpha;$$

where Δ is the critical value with the sample statistic shall be compared. Using Table C.3 and for a significance level of 5% it is observed that $\Delta = 5.9915$, a value that is larger than $\varepsilon_m^2 = 0.41$. Hence the null hypothesis cannot be rejected at the 5% significance level.

5.12 An engineer wants to examine and compare the suitability of two distribution function model alternatives for a random material property. Measurements are taken of the material property. The engineer uses the two model alternatives to calculate the Chi-square sample statistics and the corresponding sample likelihoods. The results are given in the following table:

Model	Degrees of freedom	Chi-square sample statistic	Sample likelihood
1	2	0.410	0.815
2	1	0.407	0.524

Which of the following statement(s) is(are) correct?
The engineer may accept model 1 at the 5% significance level. ✓
The engineer may accept model 2 at the 5% significance level. ✓
Model 1 is more suitable than model 2. ✓
None of the above. □

A.6 Chapter 6

6.1 It is convenient to describe failure events in terms of functional relations. If fulfilled, the considered event will occur. A failure event may be described by a functional relation, the limit state function $g(\mathbf{x})$, such as: $F = \{g(\mathbf{x}) \leq 0\}$, where the components of the vector $\mathbf{x}$ are realizations of the so-called basic random variables $\mathbf{X}$ representing all the relevant uncertainties influencing the probability of failure (*see also Sect.* 6.2).

6.2 The reliability index may be defined as the shortest distance between the curve represented by the limit state function and the origin. The reliability index β is related to the probability of failure P_F as: $P_F = \Phi(-\beta)$ (*see also Sect.* 6.3).

6.3 The estimate of the failure probability becomes exact as the number of simulation approaches infinity (*see also Sect.* 6.6).

6.4 The reliability index β can be calculated by Eq. 6.10, Sect. 6.3: $\beta = \frac{\mu_M}{\sigma_M}$. The mean μ_M and standard deviation σ_M of the safety margin $M = R - L$ can be calculated by applying the properties of the expectation operator (*see Sect.* 4.3) on the safety margin expression. This gives:

$$\mu_M = \mu_R - \mu_L = 30 - 9 = 21 \text{ kNm} \quad \text{and}$$

$$\sigma_M = \sqrt{\sigma_R^2 + \sigma_L^2} = \sqrt{5^2 + 2^2} = 5.39 \text{ kNm}$$

Hence the reliability index is equal to:

$$\beta = \frac{\mu_M}{\sigma_M} = \frac{21}{5.39} = 3.9$$

The annual probability of failure of the timber beam is:

$$P_F = P(M \leq 0) = \Phi(-\beta) = \Phi(-3.9) = 4.8 \cdot 10^{-5}$$

where $\Phi(-3.9) = 4.8 \cdot 10^{-5}$, can be found from Table C.1.

6.5 The safety margin can be written as:

$$M = R - S = A \cdot f_y - S = 100 \cdot f_y - 35$$

Since the yield stress f_y is Normal distributed, M is also Normal distributed and its mean and standard deviation can be calculated as follows:

$$\mu_M = E[M] = E[100 \cdot f_y - 35] = 100 \cdot \mu_{f_y} - 35$$
$$= 100 \cdot 425 \cdot 10^{-3} - 35 = 7.5 \text{ kN}$$

The variance is calculated as:

$$\sigma_M^2 = Var[M] = Var[100 \cdot f_y - 35] = Var[100 \cdot f_y] - Var[35] =$$
$$= 100^2 \cdot \sigma_{f_y}^2 - 0 = 100^2 \cdot (25 \cdot 10^{-3})^2 = 6.25 \text{ kN}^2$$

The standard deviation is then:

$$\sigma_M = \sqrt{\sigma_M^2} = \sqrt{6.25} = 2.5 \text{ kN}$$

The probability of failure of the rod is then (*following Eq.* 6.8, *Sect.* 6.3):

$$P_f = P(M \leq 0) = P\left(Z_M \leq \frac{0-\mu_M}{\sigma_M}\right) = \Phi\left(\frac{0-\mu_M}{\sigma_M}\right) = \Phi\left(\frac{0-7.5}{2.5}\right)$$
$$= \Phi(-3) = 0.00135$$

Whereas the reliability of the rod is simply:

$$\text{Reliability} = 1 - P_f = 1 - 0.00135 = 0.99865$$

(Note: The standard Normal distribution value corresponding to -3 is taken from Table C.1. It is easier to draw the probability density function of the standardized safety margin i.e. of Z_M. The area under the density function to the right of -3 in the x-axis represents the safe region.)

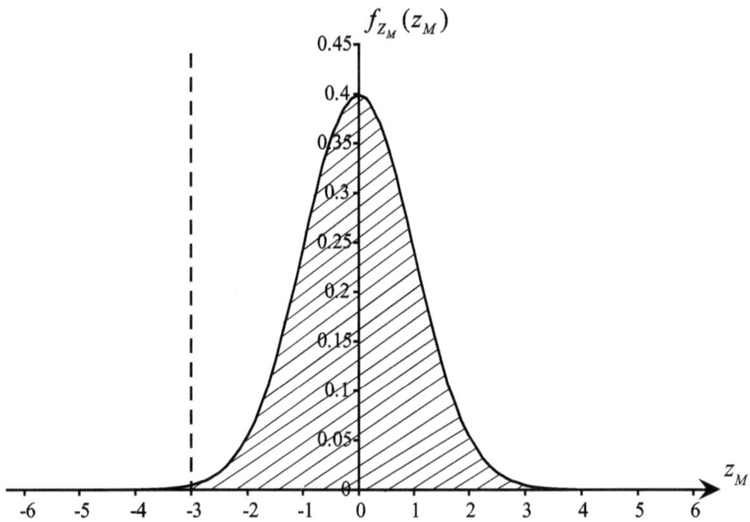

6.6 Using Fig. 6.5 and basic principles of geometry,

$$\frac{b}{\sin(\beta)} = \frac{c}{\sin(\pi - \alpha - \beta)} \quad \Rightarrow \quad b = f(c,\ \alpha,\ \beta) = c \cdot \frac{\sin(\beta)}{\sin(\alpha + \beta)}$$

Using the properties of the expectation operator (*see Sect.* 4.3),

$$E[b] = E\left[c \cdot \frac{\sin(\beta)}{\sin(\alpha+\beta)}\right] = E[c] \cdot \frac{\sin(\beta)}{\sin(\alpha+\beta)}$$
$$= 6 \cdot \frac{\sin(1.225)}{\sin(1.225 + 0.813)} = 6.32 \text{ km}$$

The error associated with the measurement of side is represented by the standard deviation $\sigma[b]$ and is estimated as:

$$V[b] = \left[\frac{\partial f}{\partial c}\right]^2 \cdot {\sigma_c}^2 + \left[\frac{\partial f}{\partial \alpha}\right]^2 \cdot {\sigma_\alpha}^2 + \left[\frac{\partial f}{\partial \beta}\right]^2 \cdot {\sigma_\beta}^2$$

$$\frac{\partial f}{\partial c} = \frac{\sin(\beta)}{\sin(\alpha+\beta)}$$

$$\frac{\partial f}{\partial \alpha} = \frac{\partial}{\partial \alpha}\left(c \cdot \frac{\sin(\beta)}{\sin(\alpha+\beta)}\right) = \frac{\partial}{\partial \alpha}\left(c \cdot \sin(\beta) \cdot \sin(\alpha+\beta)^{-1}\right)$$

$$= -c \cdot \frac{\sin(\beta) \cdot \cos(\alpha+\beta)}{(\sin(\alpha+\beta))^2}$$

$$\frac{\partial f}{\partial \beta} = \frac{\partial}{\partial \beta}\left(c \cdot \frac{\sin(\beta)}{\sin(\alpha+\beta)}\right) = \frac{\partial}{\partial \beta}\left(c \cdot \sin(\beta) \cdot (\sin(\alpha+\beta))^{-1}\right)$$

$$= \big(c \cdot \cos(\beta) \cdot (\sin(\alpha+\beta))^{-1}$$

$$+ (-1) \cdot (\sin(\alpha+\beta))^{-2} \cdot \cos(\alpha+\beta) \cdot (1) \cdot c \cdot \sin(\beta)\big)$$

$$= c \cdot \left(\frac{\cos(\beta)}{\sin(\alpha+\beta)} - \frac{\sin(\beta) \cdot \cos(\alpha+\beta)}{(\sin(\alpha+\beta))^2}\right)$$

$$= c \cdot \left(\frac{\cos(\beta) \cdot \sin(\alpha+\beta) - \sin(\beta) \cdot \cos(\alpha+\beta)}{(\sin(\alpha+\beta))^2}\right)$$

$$= c \cdot \left(\frac{\sin(\alpha+\beta-\beta)}{(\sin(\alpha+\beta))^2}\right) = c \cdot \left(\frac{\sin(\alpha)}{(\sin(\alpha+\beta))^2}\right)$$

And eventually:

$$V[b] = \left[\frac{\partial f}{\partial c}\right]^2 \cdot \sigma_c^2 + \left[\frac{\partial f}{\partial \alpha}\right]^2 \cdot \sigma_\alpha^2 + \left[\frac{\partial f}{\partial \beta}\right]^2 \cdot \sigma_\beta^2$$

$$= \left[\frac{\sin\beta}{\sin(\alpha+\beta)}\right]^2 \cdot \sigma_c^2 + \left[\frac{c \cdot \sin\beta \cdot \cos(\alpha+\beta)}{(\sin(\alpha+\beta))^2}\right]^2 \cdot \sigma_\alpha^2$$

$$+ \left[\frac{c \cdot \sin\alpha}{(\sin(\alpha+\beta))^2}\right]^2 \sigma_\beta^2$$

$$= 1.0537^2 \cdot 0.005^2 + 3.1894^2 \cdot 0.011^2 + 5.4671^2 \cdot 0.011^2$$

$$= 0.004875\ \text{km}^2$$

The error in b is calculated by:

$$\sigma[b] = \sqrt{0.004875} = 0.0698 \text{ km}$$

A.7 Chapter 7

7.1 The probabilistic models concerning events of interest and the consequences for each event and action.

7.2 Utility function is a numerical assessment of the preferences of the decision maker (*see also Sect.* 7.3).

7.3 Prior decision analysis is based on existing information and experience for a first estimate of the probability of the considered events. In posterior decision analysis new information is used to update the above probabilities and carry out a reassessment of the decision problem (*see also Sects.* 7.6 *and* 7.7).

7.4 In pre-posterior decision analysis the decision maker can evaluate whether it is useful or not to "buy" new information that will enable to make her/his final decision (*see also Sect.* 7.8).

7.5 Using the information provided,

$$P(SN) = 0.6$$

$$P(ST) = 0.4$$

$$P(I_{ST}|ST) = 0.75$$

$$P(I_{SN}|SN) = 0.75$$

$$P(I_{ST}|SN) = 1 - P(I_{ST}|ST) = 1 - 0.75 = 0.25$$

$$P(I_{SN}|ST) = 1 - P(I_{SN}|SN) = 1 - 0.75 = 0.25$$

Using the Bayes' rule,

$$P(ST|I_{ST}) = \frac{P(I_{ST}|ST) \cdot P(ST)}{P(I_{ST}|ST) \cdot P(ST) + P(I_{ST}|SN) \cdot P(SN)}$$

$$= \frac{0.75 \cdot 0.4}{0.75 \cdot 0.4 + 0.25 \cdot 0.6} = 0.6667$$

$$P(ST|I_{SN}) = \frac{P(I_{SN}|ST) \cdot P(ST)}{P(I_{SN}|ST) \cdot P(ST) + P(I_{SN}|SN) \cdot P(SN)}$$

$$= \frac{0.25 \cdot 0.4}{0.25 \cdot 0.4 + 0.75 \cdot 0.6}$$

$$= \frac{2}{11} = 0.18182$$

$$
\begin{aligned}
P(SN|I_{ST}) &= \frac{P(I_{ST}|SN)\cdot P(SN)}{P(I_{ST}|SN)\cdot P(SN)+P(I_{ST}|ST)\cdot P(ST)}\\
&= \frac{0.25\cdot 0.6}{0.25\cdot 0.6+0.75\cdot 0.4}\\
&= \frac{1}{3} = 0.3333
\end{aligned}
$$

$$
\begin{aligned}
P(SN|I_{SN}) &= \frac{P(I_{SN}|SN)\cdot P(SN)}{P(I_{SN}|SN)\cdot P(SN)+P(I_{SN}|ST)\cdot P(ST)}\\
&= \frac{0.75\cdot 0.6}{0.75\cdot 0.6+0.25\cdot 0.4}\\
&= \frac{9}{11} = 0.8181
\end{aligned}
$$

$$
\begin{aligned}
P(I_{SN}) &= P(I_{SN}|SN)\cdot P\,(SN)+P(I_{SN}|SD)\cdot P\,(SD)\\
&= 0.75\cdot 0.6+0.25\cdot 0.4 = 0.55\\
P(I_{ST}) &= P(I_{ST}|SN)\cdot P\,(SN)+P(I_{ST}|ST)\cdot P\,(ST)\\
&= 0.25\cdot 0.6+0.75\cdot 0.4 = 0.45
\end{aligned}
$$

The event tree can now be completed. An example of calculation is provided in the following. Consider the branch associated with the activity "clean up the roof". If the roof is cleaned up there are two events that may occur according to the problem:

- the roof may collapse (due to various reasons)
- the roof will not collapse (survival of the roof)

These events are associated with some probability as shown in the event tree branches:

$$
P_f(OS) = 5\cdot 10^{-4} \quad \text{and} \quad P_s(OS) = 1-5\cdot 10^{-4} = 0.9995.
$$

Hence the expected cost of this action is:

$$
\begin{aligned}
E[C_{\text{clean up}}] &= P_f(OS)\cdot 1004000+P_s(OS)\cdot 4000\\
&= 5\cdot 10^{-4}\cdot 1004000+0.9995\cdot 4000\\
&= 4500 \text{ SFr}
\end{aligned}
$$

In a similar way the rest of the event tree may be completed.

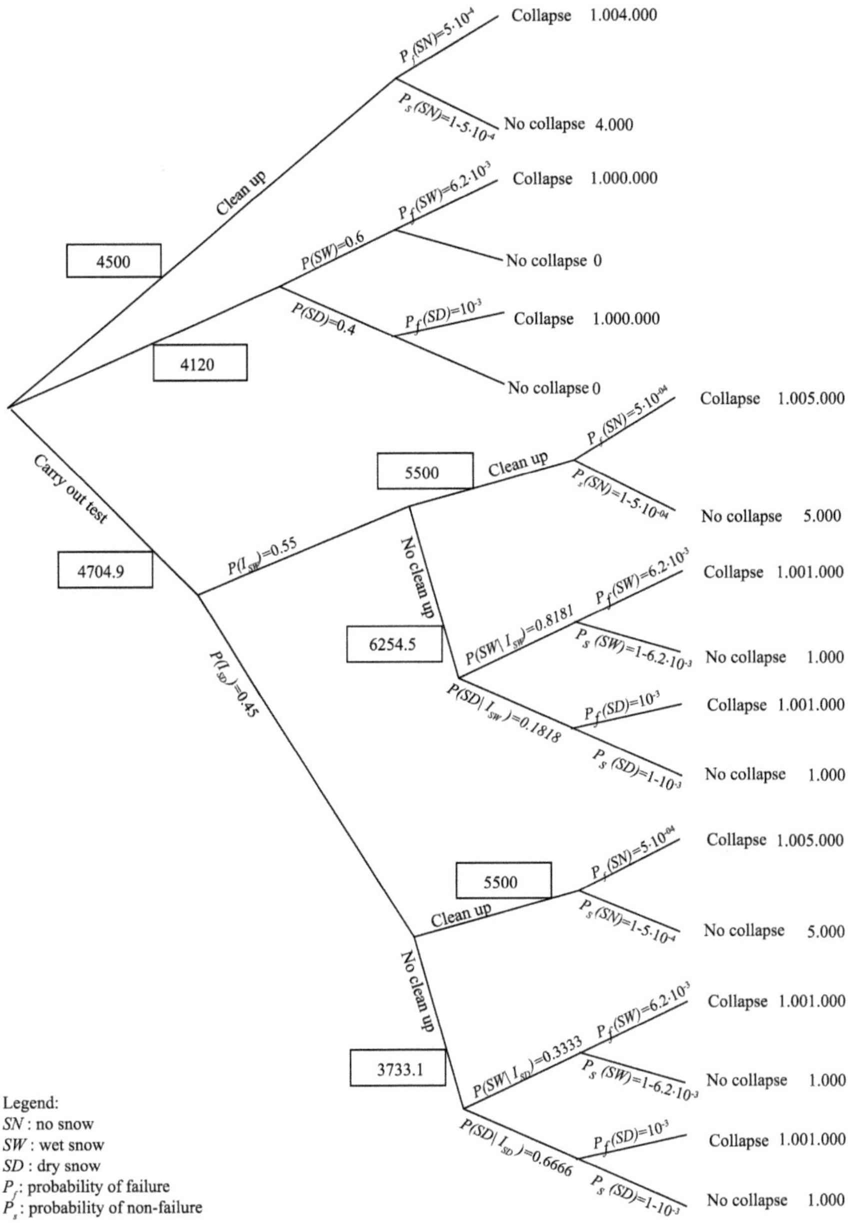

It can be seen that the action associated with the smaller cost is not to clean up the roof.

Appendix B
Examples of Calculations

B.1 Chapter 5

B.1.1 Equation 5.67

Calculation of the standard Normal distribution $\Phi(\cdot)$:

If Z is a standard Normal distributed random variable then: $p = P(Z \leq z) = \Phi(z)$. If for example $z = 0.2$ from Table C.1 it is: $\Phi(0.2) = 0.5793$.

Calculation of the inverse standard Normal distribution $\Phi^{-1}(\cdot)$:

If Z is a standard Normal distributed random variable then: $\Phi^{-1}(p) = z$. If for example $p = 0.5793$ from Table C.1 it is: $\Phi^{-1}(0.5793) = 0.2$.

So in Eq. 5.68 if the significance level α is assumed equal to 10% then it is:

$$k_{\alpha/2} = \Phi^{-1}\left(1 - \frac{\alpha}{2}\right) = \Phi^{-1}\left(1 - \frac{0.1}{2}\right) = \Phi^{-1}\,(0.95)$$

which from Table C.1, as explained above, yields (approximately): $\Phi^{-1}\,(0.95) = 1.65$ (1.645 approximately if you carry out interpolation).

B.1.2 Equation 5.71

From Eq. 5.70 the Normal distributed random variable $\bar{X}$ is transformed to a standard Normal distributed random variable such as:

$$P\left(0.3 - \Delta \leq \bar{X} \leq 0.3 + \Delta\right) = 1 - \alpha \Rightarrow$$

$$P\left(\frac{0.3 - \Delta - \mu_{\bar{X}}}{\sigma_{\bar{X}}} \leq \frac{\bar{X} - \mu_{\bar{X}}}{\sigma_{\bar{X}}} \leq \frac{0.3 + \Delta - \mu_{\bar{X}}}{\sigma_{\bar{X}}}\right) = 1 - \alpha \Rightarrow$$

$$P\left(\frac{0.3 - \Delta - \mu_{X}}{\sigma_X/\sqrt{n}} \leq \frac{\bar{X} - \mu_{\bar{X}}}{\sigma_X/\sqrt{n}} \leq \frac{0.3 + \Delta - \mu_{\bar{X}}}{\sigma_X/\sqrt{n}}\right) = 1 - \alpha \Rightarrow$$

M.H. Faber, *Statistics and Probability Theory*,
Topics in Safety, Risk, Reliability and Quality 18,
DOI 10.1007/978-94-007-4056-3, © Springer Science+Business Media B.V. 2012

$$P\left(\frac{0.3-\Delta-0.3}{0.04/\sqrt{10}} \leq \frac{\bar{X}-0.3}{0.04/\sqrt{10}} \leq \frac{0.3+\Delta-0.3}{0.04/\sqrt{10}}\right) = 1-0.1 \Rightarrow$$

$$P\left(\frac{-\Delta}{0.04/\sqrt{10}} \leq \frac{\bar{X}-0.3}{0.04/\sqrt{10}} \leq \frac{\Delta}{0.04/\sqrt{10}}\right) = 0.9$$

Following the standardization of $\bar{X}$ a new variable Z results

$$Z = \frac{\bar{X}-\mu_{\bar{X}}}{\sigma_{\bar{X}}} = \frac{\bar{X}-\mu_X}{\sigma_X/\sqrt{n}}$$

that is standard Normal distributed with mean equal to 0 and standard deviation equal to 1. So form the above equation we can write:

$$P\left(\frac{-\Delta}{0.04/\sqrt{10}} \leq \frac{\bar{X}-0.3}{0.04/\sqrt{10}} \leq \frac{\Delta}{0.04/\sqrt{10}}\right) = 0.9 \Rightarrow$$

$$P\left(Z \leq \frac{\Delta}{0.04/\sqrt{10}}\right) - P\left(Z \geq \frac{-\Delta}{0.04/\sqrt{10}}\right) = 0.9 \Rightarrow$$

$$\Phi\left(\frac{\Delta}{0.04/\sqrt{10}}\right) - \left(1-\Phi\left(\frac{\Delta}{0.04/\sqrt{10}}\right)\right) = 0.9 \Rightarrow$$

$$2\Phi\left(\frac{\Delta}{0.04/\sqrt{10}}\right) - 1 = 0.9 \Rightarrow$$

$$\Phi\left(\frac{\Delta}{0.04/\sqrt{10}}\right) = 0.95 \Rightarrow \frac{\Delta}{0.04/\sqrt{10}} = \Phi^{-1}(0.95) \Rightarrow$$

$$\Delta = \frac{0.04}{\sqrt{10}}1.645 = 0.0208$$

where the value of $\Phi^{-1}(0.95) = 1.645$ is found with the help of Table C.1.

B.1.3 Examples on Chi-Square Significance Test

The degrees of freedom in the Chi-square significance test are defined as:

$$k-1-l$$

where k is the number of intervals to which the available data are arranged and l is the number of the parameters of the assumed distribution which are calculated from the data directly.

In the example in Sect. 5.9.1, first both the distribution (Normal) and the parameters (mean and standard deviation) are postulated i.e. assumed for the concrete compressive strength data. This means that $l = 0$ since no parameter is calculated from the data.

The data are arranged into 3 intervals and so $k = 3$. Following the above expression for the degrees of freedom the sample statistic (see Eq. 5.77) has $k - 1 - l = 3 - 1 - 0 = 2$ degrees of freedom.

Subsequently, for the same example, the distribution (Normal) and the mean value are postulated i.e. assumed while the standard deviation (parameter of the assumed distribution) is calculated from the data and hence $l = 1$. So in that case, the degrees of freedom of the sample statistic are: $k - 1 - l = 3 - 1 - 1 = 1$.

The critical value Δ can be found from Table C.3 using the degrees of freedom and the probability

$$P(\varepsilon_m^2 \leq \Delta) = 1 - \alpha \left(\Leftrightarrow P(\varepsilon_m^2 \geq \Delta) = \alpha \right)$$

B.2 Chapter 6

B.2.1 Example 6.2

The variance of c is estimated as follows:

$$\begin{aligned} Var[c] &= \sum_{i=1}^{n} \left(\left. \frac{\partial h(\mathbf{x})}{\partial x_i} \right|_{\mathbf{x}=\mathbf{x}_0} \right)^2 \sigma_{X_i}^2 \\ &= \left(\left. \frac{\partial \left(\sqrt{a^2+b^2} \right)}{\partial a} \right|_{\substack{a=\eta_a, \\ b=\eta_b}} \right)^2 \sigma_\alpha^2 + \left(\left. \frac{\partial \left(\sqrt{a^2+b^2} \right)}{\partial b} \right|_{\substack{a=\eta_a, \\ b=\eta_b}} \right)^2 \sigma_b^2 \\ &= \left(\frac{2\mu_a}{2\sqrt{\mu_a^2+\mu_b^2}} \right)^2 \sigma_a^2 + \left(\frac{2\mu_b}{2\sqrt{\mu_a^2+\mu_b^2}} \right)^2 \sigma_b^2 = \frac{\mu_a^2}{\mu_a^2+\mu_b^2}\sigma_a^2 + \frac{\mu_b^2}{\mu_a^2+\mu_b^2}\sigma_b^2 \end{aligned}$$

B.2.2 Example 6.3

The reliability index is obtained by solving $g(u) = 0$ to β. $g(u)$ is already given in the example:

$$g(u) = 350u_R + 305u_A - 300u_S + 35u_R u_A + 2000$$

This can be re-written as:

$$g(u) = g(\beta\boldsymbol{\alpha}) = 350\beta\alpha_R + 35\beta\alpha_A - 300\beta\alpha_S + 35\beta^2\alpha_R\alpha_A + 2000$$

And from Eq. 6.18:

$$g(u) = 0 \Rightarrow 350\alpha_R\beta + 305\alpha_A\beta - 300\alpha_S\beta + 35\alpha_R\alpha_A\beta^2 + 2000 = 0 \Rightarrow$$

$$\beta = \frac{-2000}{350\alpha_R + 305\alpha_A - 300\alpha_S + 35\beta\alpha_R\alpha_A}$$

From Eq. 6.17:

$$\alpha_R = \frac{-\frac{\partial g}{\partial u_R}(\beta\alpha)}{\sqrt{\sum_{i=1}^{n}\left(\frac{\partial g}{\partial u_i}(\beta\alpha)\right)^2}}$$

$$\frac{\partial g}{\partial u_R}(350\beta\alpha_R + 35\beta\alpha_A - 300\beta\alpha_S + 35\beta^2\alpha_R\alpha_A + 2000) = 350 + 35\beta\alpha_A$$

$$\frac{\partial g}{\partial u_A}(350\beta\alpha_R + 35\beta\alpha_A - 300\beta\alpha_S + 35\beta^2\alpha_R\alpha_A + 2000) = 350 + 35\beta\alpha_R$$

$$\frac{\partial g}{\partial u_S}(350\beta\alpha_R + 35\beta\alpha_A - 300\beta\alpha_S + 35\beta^2\alpha_R\alpha_A + 2000) = 300$$

Hence:

$$\alpha_R = \frac{-\frac{\partial g}{\partial u_R}(\beta\alpha)}{\sqrt{\sum_{i=1}^{n}\left(\frac{\partial g}{\partial u_i}(\beta\alpha)\right)^2}} = -\frac{350 + 35\beta\alpha_A}{k}$$

where

$$k = \sqrt{(350 + 35\beta\alpha_A)^2 + (350 + 35\beta\alpha_R)^2 + 300^2}$$

Similarly α_A and α_S may be found.

The aim is to calculate β so that it represents the smallest distance to the origin. The starting values of β, α_R, α_A and α_S are assumed (see Table 6.1, Column "Start"). The next step (column for iteration 1) is calculated as follows:

$$\beta = \frac{-2000}{350\alpha_R + 305\alpha_A - 300\alpha_S + 35\beta\alpha_R\alpha_A}$$

$$= \frac{-2000}{350(-0.58) + 305(-0.58) - 300(0.58) + 35\cdot 3(-0.58)(-0.58)} = 3.6719$$

$$\alpha_R = -\frac{350 + 35\cdot 3(-0.58)}{k} = -0.5701$$

$$\alpha_R = -\frac{350 + 35\cdot 3(-0.58)}{k} = -0.5701$$

$$\alpha_R = -\frac{300}{k} = 0.5916$$

where $k = \sqrt{(350 + 35\cdot 3(-0.58))^2 + (350 + 35\cdot 3(-0.58))^2 + 300^2}$.

The iterations are continued until the value of β converges to one value (in this example convergence occurs after the 4th iteration).

Appendix C
Tables

Table C.1 Cumulative distribution function of the standard normal distribution $\Phi(z)$

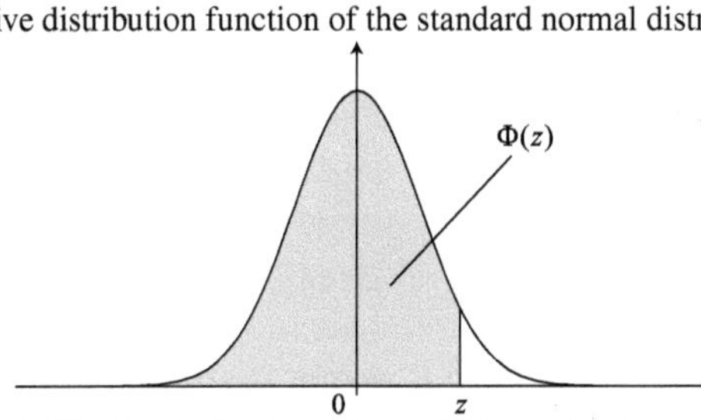

Probability density function of the standard normal random variable.

z	$\Phi(z)$	z	$\Phi(z)$	z	$\Phi(z)$	z	$\Phi(z)$	z	$\Phi(z)$
0.00	0.5000	0.50	0.6915	1.00	0.8413	1.50	0.9332	2.00	0.9772
0.01	0.5040	0.51	0.6950	1.01	0.8438	1.51	0.9345	2.10	0.9821356
0.02	0.5080	0.52	0.6985	1.02	0.8461	1.52	0.9357	2.20	0.9860966
0.03	0.5120	0.53	0.7019	1.03	0.8485	1.53	0.9370	2.30	0.9892759
0.04	0.5160	0.54	0.7054	1.04	0.8508	1.54	0.9382	2.40	0.9918025
0.05	0.5199	0.55	0.7088	1.05	0.8531	1.55	0.9394	2.50	0.9937903
0.06	0.5239	0.56	0.7123	1.06	0.8554	1.56	0.9406	2.60	0.9953388
0.07	0.5279	0.57	0.7157	1.07	0.8577	1.57	0.9418	2.70	0.9965330
0.08	0.5319	0.58	0.7190	1.08	0.8599	1.58	0.9429	2.80	0.9974449
0.09	0.5359	0.59	0.7224	1.09	0.8621	1.59	0.9441	2.90	0.9981342
0.10	0.5398	0.60	0.7257	1.10	0.8643	1.60	0.9452	3.00	0.9986501
0.11	0.5438	0.61	0.7291	1.11	0.8665	1.61	0.9463	3.10	0.9990324
0.12	0.5478	0.62	0.7324	1.12	0.8686	1.62	0.9474	3.20	0.9993129
0.13	0.5517	0.63	0.7357	1.13	0.8708	1.63	0.9484	3.30	0.9995166
0.14	0.5557	0.64	0.7389	1.14	0.8729	1.64	0.9495	3.40	0.9996631

(continued on the next page)

M.H. Faber, *Statistics and Probability Theory*,
Topics in Safety, Risk, Reliability and Quality 18,
DOI 10.1007/978-94-007-4056-3, © Springer Science+Business Media B.V. 2012

Table C.1 (Continued)

z	$\Phi(z)$	z	$\Phi(z)$	z	$\Phi(z)$	z	$\Phi(z)$	z	$\Phi(z)$
0.15	0.5596	0.65	0.7422	1.15	0.8749	1.65	0.9505	3.50	0.9997674
0.16	0.5636	0.66	0.7454	1.16	0.8770	1.66	0.9515	3.60	0.9998409
0.17	0.5675	0.67	0.7486	1.17	0.8790	1.67	0.9525	3.70	0.9998922
0.18	0.5714	0.68	0.7517	1.18	0.8810	1.68	0.9535	3.80	0.9999277
0.19	0.5753	0.69	0.7549	1.19	0.8830	1.69	0.9545	3.90	0.9999519
0.20	0.5793	0.70	0.7580	1.20	0.8849	1.70	0.9554	4.00	0.9999683
0.21	0.5832	0.71	0.7611	1.21	0.8869	1.71	0.9564	4.10	0.9999793
0.22	0.5871	0.72	0.7642	1.22	0.8888	1.72	0.9573	4.20	0.9999867
0.23	0.5910	0.73	0.7673	1.23	0.8907	1.73	0.9582	4.30	0.9999915
0.24	0.5948	0.74	0.7704	1.24	0.8925	1.74	0.9591	4.40	0.9999946
0.25	0.5987	0.75	0.7734	1.25	0.8944	1.75	0.9599	4.50	0.9999966
0.26	0.6026	0.76	0.7764	1.26	0.8962	1.76	0.9608	4.60	0.9999979
0.27	0.6064	0.77	0.7794	1.27	0.8980	1.77	0.9616	4.70	0.9999987
0.28	0.6103	0.78	0.7823	1.28	0.8997	1.78	0.9625	4.80	0.9999992
0.29	0.6141	0.79	0.7852	1.29	0.9015	1.79	0.9633	4.90	0.9999995
0.30	0.6179	0.80	0.7881	1.30	0.9032	1.80	0.9641	5.00	0.9999997
0.31	0.6217	0.81	0.7910	1.31	0.9049	1.81	0.9649		
0.32	0.6255	0.82	0.7939	1.32	0.9066	1.82	0.9656		
0.33	0.6293	0.83	0.7967	1.33	0.9082	1.83	0.9664		
0.34	0.6331	0.84	0.7995	1.34	0.9099	1.84	0.9671		
0.35	0.6368	0.85	0.8023	1.35	0.9115	1.85	0.9678		
0.36	0.6406	0.86	0.8051	1.36	0.9131	1.86	0.9686		
0.37	0.6443	0.87	0.8078	1.37	0.9147	1.87	0.9693		
0.38	0.6480	0.88	0.8106	1.38	0.9162	1.88	0.9699		
0.39	0.6517	0.89	0.8133	1.39	0.9177	1.89	0.9706		
0.40	0.6554	0.90	0.8159	1.40	0.9192	1.90	0.9713		
0.41	0.6591	0.91	0.8186	1.41	0.9207	1.91	0.9719		
0.42	0.6628	0.92	0.8212	1.42	0.9222	1.92	0.9726		
0.43	0.6664	0.93	0.8238	1.43	0.9236	1.93	0.9732		
0.44	0.6700	0.94	0.8264	1.44	0.9251	1.94	0.9738		
0.45	0.6736	0.95	0.8289	1.45	0.9265	1.95	0.9744		
0.46	0.6772	0.96	0.8315	1.46	0.9279	1.96	0.9750		
0.47	0.6808	0.97	0.8340	1.47	0.9292	1.97	0.9756		
0.48	0.6844	0.98	0.8365	1.48	0.9306	1.98	0.9761		
0.49	0.6879	0.99	0.8389	1.49	0.9319	1.99	0.9767		

Table C.2 Quantiles q of the t-distribution

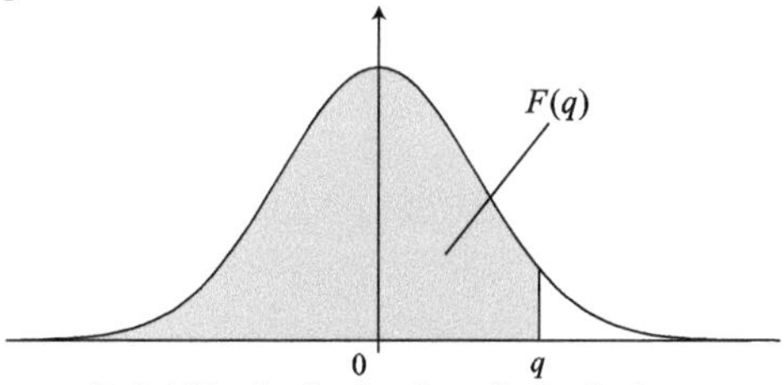

Probability density function of t-distribution

ν^a	$F(q)$							
	0.75	0.8	0.85	0.9	0.95	0.975	0.99	0.995
1	1.000	1.376	1.963	3.078	6.314	12.706	31.821	63.657
2	0.816	1.061	1.386	1.886	2.920	4.303	6.965	9.925
3	0.765	0.978	1.250	1.638	2.353	3.182	4.541	5.841
4	0.741	0.941	1.190	1.533	2.132	2.776	3.747	4.604
5	0.727	0.920	1.156	1.476	2.015	2.571	3.365	4.032
6	0.718	0.906	1.134	1.440	1.943	2.447	3.143	3.707
7	0.711	0.896	1.119	1.415	1.895	2.365	2.998	3.499
8	0.706	0.889	1.108	1.397	1.860	2.306	2.896	3.355
9	0.703	0.883	1.100	1.383	1.833	2.262	2.821	3.250
10	0.700	0.879	1.093	1.372	1.812	2.228	2.764	3.169
11	0.697	0.876	1.088	1.363	1.796	2.201	2.718	3.106
12	0.695	0.873	1.083	1.356	1.782	2.179	2.681	3.055
13	0.694	0.870	1.079	1.350	1.771	2.160	2.650	3.012
14	0.692	0.868	1.076	1.345	1.761	2.145	2.624	2.977
15	0.691	0.866	1.074	1.341	1.753	2.131	2.602	2.947
16	0.690	0.865	1.071	1.337	1.746	2.120	2.583	2.921
17	0.689	0.863	1.069	1.333	1.740	2.110	2.567	2.898
18	0.688	0.862	1.067	1.330	1.734	2.101	2.552	2.878
19	0.688	0.861	1.066	1.328	1.729	2.093	2.539	2.861
20	0.687	0.860	1.064	1.325	1.725	2.086	2.528	2.845
21	0.686	0.859	1.063	1.323	1.721	2.080	2.518	2.831
22	0.686	0.858	1.061	1.321	1.717	2.074	2.508	2.819
23	0.685	0.858	1.060	1.319	1.714	2.069	2.500	2.807
24	0.685	0.857	1.059	1.318	1.711	2.064	2.492	2.797
25	0.684	0.856	1.058	1.316	1.708	2.060	2.485	2.787
26	0.684	0.856	1.058	1.315	1.706	2.056	2.479	2.779
27	0.684	0.855	1.057	1.314	1.703	2.052	2.473	2.771
28	0.683	0.855	1.056	1.313	1.701	2.048	2.467	2.763
29	0.683	0.854	1.055	1.311	1.699	2.045	2.462	2.756

(*continued on the next page*)

Table C.2 (Continued)

ν^{a}	$F(q)$ 0.75	0.8	0.85	0.9	0.95	0.975	0.99	0.995
30	0.683	0.854	1.055	1.310	1.697	2.042	2.457	2.750
40	0.681	0.851	1.050	1.303	1.684	2.021	2.423	2.704
50	0.679	0.849	1.047	1.299	1.676	2.009	2.403	2.678
60	0.679	0.848	1.045	1.296	1.671	2.000	2.390	2.660
70	0.678	0.847	1.044	1.294	1.667	1.994	2.381	2.648
80	0.678	0.846	1.043	1.292	1.664	1.990	2.374	2.639
90	0.677	0.846	1.042	1.291	1.662	1.987	2.368	2.632
100	0.677	0.845	1.042	1.290	1.660	1.984	2.364	2.626
∞	0.674	0.842	1.036	1.282	1.645	1.960	2.326	2.576

[a]Degrees of freedom.

Table C.3 Quantiles q of the Chi-square distribution

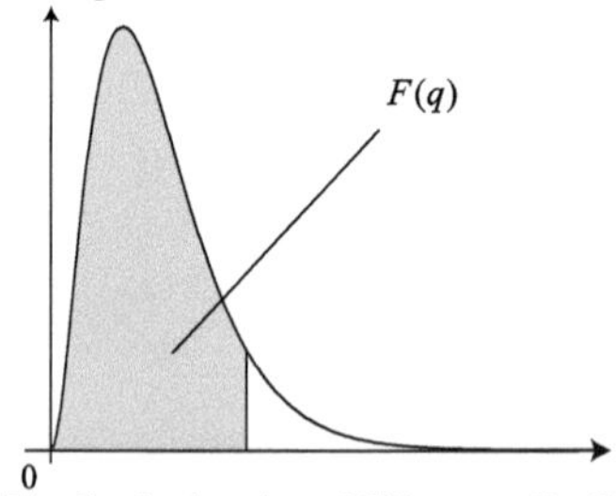

Probability density function of Chi-square distribution

ν^{a}	$F(q)$ 0.75	0.90	0.95	0.98	0.99	0.995	0.999
1	1.3233	2.7055	3.8415	5.4119	6.6349	7.8794	10.8276
2	2.7726	4.6052	5.9915	7.8240	9.2103	10.5966	13.8155
3	4.1083	6.2514	7.8147	9.8374	11.3449	12.8382	16.2662
4	5.3853	7.7794	9.4877	11.6678	13.2767	14.8603	18.4668
5	6.6257	9.2364	11.0705	13.3882	15.0863	16.7496	20.5150
6	7.8408	10.6446	12.5916	15.0332	16.8119	18.5476	22.4577
7	9.0371	12.0170	14.0671	16.6224	18.4753	20.2777	24.3219
8	10.2189	13.3616	15.5073	18.1682	20.0902	21.9550	26.1245
9	11.3888	14.6837	16.9190	19.6790	21.6660	23.5894	27.8772
10	12.5489	15.9872	18.3070	21.1608	23.2093	25.1882	29.5883
11	13.7007	17.2750	19.6751	22.6179	24.7250	26.7568	31.2641
12	14.8454	18.5493	21.0261	24.0540	26.2170	28.2995	32.9095
13	15.9839	19.8119	22.3620	25.4715	27.6882	29.8195	34.5282
14	17.1169	21.0641	23.6848	26.8728	29.1412	31.3193	36.1233

(continued on the next page)

Table C.3 (Continued)

ν[a]	$F(q)$						
	0.75	0.90	0.95	0.98	0.99	0.995	0.999
15	18.2451	22.3071	24.9958	28.2595	30.5779	32.8013	37.6973
16	19.3689	23.5418	26.2962	29.6332	31.9999	34.2672	39.2524
17	20.4887	24.7690	27.5871	30.9950	33.4087	35.7185	40.7902
18	21.6049	25.9894	28.8693	32.3462	34.8053	37.1565	42.3124
19	22.7178	27.2036	30.1435	33.6874	36.1909	38.5823	43.8202
20	23.8277	28.4120	31.4104	35.0196	37.5662	39.9968	45.3147
21	24.9348	29.6151	32.6706	36.3434	38.9322	41.4011	46.7970
22	26.0393	30.8133	33.9244	37.6595	40.2894	42.7957	48.2679
23	27.1413	32.0069	35.1725	38.9683	41.6384	44.1813	49.7282
24	28.2412	33.1962	36.4150	40.2704	42.9798	45.5585	51.1786
25	29.3389	34.3816	37.6525	41.5661	44.3141	46.9279	52.6197
26	30.4346	35.5632	38.8851	42.8558	45.6417	48.2899	54.0520
27	31.5284	36.7412	40.1133	44.1400	46.9629	49.6449	55.4760
28	32.6205	37.9159	41.3371	45.4188	48.2782	50.9934	56.8923
29	33.7109	39.0875	42.5570	46.6927	49.5879	52.3356	58.3012
30	34.7997	40.2560	43.7730	47.9618	50.8922	53.6720	59.7031

[a]Degrees of freedom.

Table C.4 Critical values of the Kolmogorov-Smirnov test

n[a]	α[b]				
	0.01	0.02	0.05	0.1	0.2
1	0.995	0.990	0.975	0.950	0.900
2	0.929	0.900	0.842	0.776	0.684
3	0.829	0.785	0.708	0.636	0.565
4	0.734	0.689	0.624	0.565	0.493
5	0.669	0.627	0.563	0.509	0.447
6	0.617	0.577	0.519	0.468	0.410
7	0.576	0.538	0.483	0.436	0.381
8	0.542	0.507	0.454	0.410	0.358
9	0.513	0.480	0.430	0.387	0.339
10	0.489	0.457	0.409	0.369	0.323
11	0.468	0.437	0.391	0.352	0.308
12	0.449	0.419	0.375	0.338	0.296
13	0.432	0.404	0.361	0.325	0.285
14	0.418	0.390	0.349	0.314	0.275
15	0.404	0.377	0.338	0.304	0.266

(continued on the next page)

Table C.4 (Continued)

n[a]	α[b]				
	0.01	0.02	0.05	0.1	0.2
16	0.392	0.366	0.327	0.295	0.258
17	0.381	0.355	0.318	0.286	0.250
18	0.371	0.346	0.309	0.279	0.244
19	0.361	0.337	0.301	0.271	0.237
20	0.352	0.329	0.294	0.265	0.232
21	0.344	0.321	0.287	0.259	0.226
22	0.337	0.314	0.281	0.253	0.221
23	0.330	0.307	0.275	0.248	0.217
24	0.323	0.301	0.269	0.242	0.212
25	0.317	0.295	0.264	0.238	0.208
26	0.311	0.290	0.259	0.233	0.204
27	0.305	0.284	0.254	0.229	0.200
28	0.300	0.279	0.250	0.225	0.197
29	0.295	0.275	0.246	0.221	0.194
30	0.290	0.270	0.242	0.218	0.190
31	0.285	0.266	0.238	0.214	0.187
32	0.281	0.262	0.234	0.211	0.185
33	0.277	0.258	0.231	0.208	0.182
34	0.273	0.254	0.227	0.205	0.179
35	0.269	0.251	0.224	0.202	0.177
36	0.265	0.247	0.221	0.199	0.174
37	0.262	0.244	0.218	0.196	0.172
38	0.258	0.241	0.215	0.194	0.170
39	0.255	0.238	0.213	0.192	0.168
40	0.252	0.235	0.210	0.189	0.166
$n > 40$	$1.63/\sqrt{n}$	$1.52/\sqrt{n}$	$1.36/\sqrt{n}$	$1.22/\sqrt{n}$	$1.07/\sqrt{n}$

[a]Sample size.
[b]Significance level.

Table C.5 Gamma function $\Gamma(x) = \int_0^\infty e^{-t} t^{x-1}\, dt$

x	$\Gamma(x)$	x	$\Gamma(x)$	x	$\Gamma(x)$	x	$\Gamma(x)$	x	$\Gamma(x)$	x	$\Gamma(x)$
0.00010	9999.4	0.0010	999.42	0.010	99.43	0.10	9.514	1.0	1.000	5.5	52.3
0.00012	8332.8	0.0012	832.76	0.012	82.77	0.12	7.863	1.1	0.951	5.6	61.6
0.00014	7142.3	0.0014	713.71	0.014	70.87	0.14	6.689	1.2	0.918	5.7	72.5
0.00016	6249.4	0.0016	624.42	0.016	61.94	0.16	5.811	1.3	0.897	5.8	85.6

(*continued on the next page*)

Table C.5 (Continued)

x	$\Gamma(x)$	x	$\Gamma(x)$	x	$\Gamma(x)$	x	$\Gamma(x)$	x	$\Gamma(x)$	x	$\Gamma(x)$
0.00018	5555.0	0.0018	554.98	0.018	55.00	0.18	5.132	1.4	0.887	5.9	101.3
0.00020	4999.4	0.0020	499.42	0.020	49.44	0.20	4.591	1.5	0.886	6.0	120.0
0.00022	4544.9	0.0022	453.97	0.022	44.90	0.22	4.150	1.6	0.894	6.1	142.5
0.00024	4166.1	0.0024	416.09	0.024	41.11	0.24	3.786	1.7	0.909	6.2	169.4
0.00026	3845.6	0.0026	384.04	0.026	37.91	0.26	3.478	1.8	0.931	6.3	201.8
0.00028	3570.9	0.0028	356.57	0.028	35.16	0.28	3.217	1.9	0.962	6.4	240.8
0.00030	3332.8	0.0030	332.76	0.030	32.78	0.30	2.992	2.0	1.000	6.5	287.9
0.00032	3124.4	0.0032	311.93	0.032	30.70	0.32	2.796	2.1	1.046	6.6	344.7
0.00034	2940.6	0.0034	293.54	0.034	28.87	0.34	2.624	2.2	1.102	6.7	413.4
0.00036	2777.2	0.0036	277.20	0.036	27.24	0.36	2.473	2.3	1.167	6.8	496.6
0.00038	2631.0	0.0038	262.58	0.038	25.77	0.38	2.338	2.4	1.242	6.9	597.5
0.00040	2499.4	0.0040	249.43	0.040	24.46	0.40	2.218	2.5	1.329	7.0	720.0
0.00042	2380.4	0.0042	237.52	0.042	23.27	0.42	2.110	2.6	1.430	7.1	869.0
0.00044	2272.2	0.0044	226.70	0.044	22.19	0.44	2.013	2.7	1.545	7.2	1050
0.00046	2173.3	0.0046	216.82	0.046	21.21	0.46	1.925	2.8	1.676	7.3	1271
0.00048	2082.8	0.0048	207.76	0.048	20.30	0.48	1.845	2.9	1.827	7.4	1541
0.00050	1999.4	0.0050	199.43	0.050	19.47	0.50	1.772	3.0	2.000	7.5	1871
0.00052	1922.5	0.0052	191.74	0.052	18.70	0.52	1.706	3.1	2.198	7.6	2275
0.00054	1851.3	0.0054	184.61	0.054	17.99	0.54	1.645	3.2	2.424	7.7	2770
0.00056	1785.1	0.0056	178.00	0.056	17.33	0.56	1.589	3.3	2.683	7.8	3377
0.00058	1723.6	0.0058	171.84	0.058	16.72	0.58	1.537	3.4	2.981	7.9	4123
0.00060	1666.1	0.0060	166.10	0.060	16.15	0.60	1.489	3.5	3.323	8.0	5040
0.00062	1612.3	0.0062	160.72	0.062	15.61	0.62	1.445	3.6	3.717	8.1	6170
0.00064	1561.9	0.0064	155.68	0.064	15.11	0.64	1.404	3.7	4.171	8.2	7562
0.00066	1514.6	0.0066	150.94	0.066	14.64	0.66	1.366	3.8	4.694	8.3	9281
0.00068	1470.0	0.0068	146.49	0.068	14.19	0.68	1.331	3.9	5.299	8.4	11406
0.00070	1428.0	0.0070	142.29	0.070	13.77	0.70	1.298	4.0	6.000	8.5	14034
0.00072	1388.3	0.0072	138.32	0.072	13.38	0.72	1.267	4.1	6.813	8.6	17290
0.00074	1350.8	0.0074	134.57	0.074	13.00	0.74	1.239	4.2	7.757	8.7	21328
0.00076	1315.2	0.0076	131.01	0.076	12.65	0.76	1.212	4.3	8.855	8.8	26340
0.00078	1281.5	0.0078	127.64	0.078	12.32	0.78	1.187	4.4	10.136	8.9	32569
0.00080	1249.4	0.0080	124.43	0.080	12.00	0.80	1.164	4.5	11.632	9.0	40320
0.00082	1218.9	0.0082	121.38	0.082	11.69	0.82	1.142	4.6	13.381	9.1	49974
0.00084	1189.9	0.0084	118.48	0.084	11.40	0.84	1.122	4.7	15.431	9.2	62011
0.00086	1162.2	0.0086	115.71	0.086	11.13	0.86	1.103	4.8	17.838	9.3	77036
0.00088	1135.8	0.0088	113.07	0.088	10.87	0.88	1.085	4.9	20.667	9.4	95809
0.00090	1110.5	0.0090	110.54	0.090	10.62	0.90	1.069	5.0	24.000	9.5	119292
0.00092	1086.4	0.0092	108.13	0.092	10.38	0.92	1.053	5.1	27.932	9.6	148696
0.00094	1063.3	0.0094	105.81	0.094	10.15	0.94	1.038	5.2	32.578	9.7	185551
0.00096	1041.1	0.0096	103.60	0.096	9.93	0.96	1.025	5.3	38.078	9.8	231792
0.00098	1019.8	0.0098	101.47	0.098	9.72	0.98	1.012	5.4	44.599	9.9	289868

References

1. Alsalam, N. et al.: Economic Effects of Federal Spending on Infrastructure and Other Investments. Congressional Budget Office, available from: http://www.cbo.gov (1998)
2. Ang, A.H., Tang, W.H.: Probability Concepts in Engineering: Emphasis on Applications in Civil & Environmental Engineering, 2nd edn. Wiley, New York (2007)
3. Basler, E.: Untersuchungen über den Sicherheitsbegriff von Bauwerken. Schweizer Archiv für angewandte Wissenschaft und Technik (1961)
4. Benjamin, J.R., Cornell, C.A.: Probability, Statistics and Decision for Civil Engineers. McGraw-Hill, New York (1971)
5. Brundtland, G.H.: Our Common Future. World Commission on Environment and Development, 1987.
6. Cornell, C.A.: A probability based structural code. ACI J. **66**, 974–985 (1969)
7. Hasofer, A.M., Lind, N.C.: An exact and invariant first order reliability format. J. Eng. Mech. Div. Proc. ASCE **100**(EMl), 111–121 (1974)
8. JCSS: Probabilistic model code. The Joint Committee on Structural Safety (2001)
9. Madsen, H.O., Krenk, S., Lind, N.C.: Methods of Structural Safety. Dover, New York (2006)
10. Melchers, R.E.: Structural Reliability Analysis and Prediction, 2nd edn. John Wiley, New York (1999)
11. Schneider, J.: Sicherheit und Zuverlässigkeit im Bauwesen, Grundwissen für Ingenieure. VDF, Hochschulverlag AG an der ETH Zürich, Zürich (1994)
12. Stewart, N.G., Melchers, R.E.: Probabilistic Risk Assessment of Engineering Systems. Chapman & Hall, London (1997)
13. Thoft-Christensen, P., Baker, M.J.: Structural Reliability Theory and Its Applications. Springer, Berlin (1982)

M.H. Faber, *Statistics and Probability Theory*,
Topics in Safety, Risk, Reliability and Quality 18,
DOI 10.1007/978-94-007-4056-3,

Index

M.H. Faber, *Statistics and Probability Theory*,
Topics in Safety, Risk, Reliability and Quality 18,
DOI 10.1007/978-94-007-4056-3, © Springer Science+Business Media B.V. 2012

MIX
Papier aus verantwortungsvollen Quellen
Paper from responsible sources
FSC® C105338

If you have any concerns about our products,
you can contact us on
ProductSafety@springernature.com

In case Publisher is established outside the EU,
the EU authorized representative is:
Springer Nature Customer Service Center GmbH
Europaplatz 3, 69115 Heidelberg, Germany

Printed by Libri Plureos GmbH
in Hamburg, Germany